I dedicate this book to my parents, wife, and children. They supported my love of science, and were patient with me as I spent untold hours working on manuscripts.

Special thanks to my engineers, doctors, and professors who spent untold hours proofreading my rough drafts, you know who you are: Chris Cusack, Dr. Warren Jesek, Katrina Fisher and Professor Dave Kirby. Because of your donation of time, this is a much better document.

TABLE OF CONTENTS

CHAPTER 1:   MYTHS AND MISCONCEPTIONS

I have been a biology educator since the fall of 1982. In the early days of my career I felt Darwin's theory had some scientific merit, but I devoted very little teaching time to the topic because uncontested evidence seemed lacking. As the years ticked by, huge technological advances were made in the fields of genetics, paleontology, microbiology, astronomy, particle physics, and biochemistry. I started noticing one common denominator appearing in all of those fields of science, and it was the underlying theme of **COMPLEXITY.** The multiple layers of complexity that are astounding today's researchers, have increasingly convinced scientists that random chance interaction of atoms could never in a million years… or even 15 billion years… produce matter, energy, and the millions of species that have called the Earth home. World renowned atheist and brilliant Oxford professor, Antony Flew, exemplifies this shift in scientific thinking. He shocked the world recently by renouncing his lifelong naturalistic *(atheistic/Darwinian)* world view. Flew, who holds no particular religious affiliation, has concluded in his new book "Has Science Discovered God?", that some type of intelligent agent had to have created the universe and life:

> *"They (Scientists) have shown, by almost unbelievable complexity of the arrangements which are needed to produce it (life), that intelligence must have been involved. A super-intelligence is the only good explanation for the origin of life and the complexity of nature."*
>   Antony Flew, Veteran Oxford/Aberdeen professor, former atheist, *"Has Science Discovered God"*, Harper-Collins Publishing, 2004.

But many other Darwinian proponents refuse to acknowledge these compelling 'intelligence' evidences, and they refuse to consider any alternative theories like "intelligent design" (ID) if the theory falls outside of the classic naturalistic world view:

> *"Even if all the data points to an intelligent designer, such a hypothesis is excluded from science because it is not naturalistic."*
>   Dr. Scott Todd., Kansas State University, *Nature*, 401 (6752): 423, Sept. 30, 1999.

Most of the naturalists I have talked to, openly agree with Dr. Todd's viewpoint and insist that intelligent design is simply a warmed-over rehash of creationism. While a few creationists do attempt to use the label of ID as a cover to put religious agendas back into public schools, most ID proponents have no particular religious affiliation, or hidden agenda. To them, ID is broadly defined as: "an unidentified intelligent agent that engineered matter and organic life". It is my professional opinion that over the past 25 years, Darwinian arguments have become scientifically weaker, while design evidences have grown stronger. Scientists are trained to follow the evidence trail no matter where it leads. We are supposed to be open-minded and flexible researchers, unlike scientist Scott Todd. Scientific educators have a duty to their students to challenge their thinking on all scientific issues, especially issues that are as fluid and unproven as Darwinian Evolution. In the public school arena, I have noticed that several myths exist concerning the teaching of Darwinism:

MYTH ONE: If a topic has both scientific and religious aspects, only the scientific aspects are legal areas of discussion. This is wrong, because the Constitution and Supreme Court have consistently ruled that as long as a specific religion is not promoted, knowledge of religious terms and concepts is a must for a complete education. The generic topic of 'creator' is a valid and legal area of scientific discussion in public school science classrooms.

MYTH TWO: Citing specific religious scriptures is the only way to argue Darwinism. This is also wrong, because a solid scientific argument can be made against Darwinism totally on the evidences of geology, micro-biology, genetics, paleontology and thermodynamics. Religious documents should never be used as evidence in public school evolution curricula.

MYTH THREE: All Darwinists are in agreement about evolutionary processes and evidences. Actually, all that they agree upon is the basic philosophy of the theory. As you are about to see, their evidence is far from proven, and it often contradicts proven scientific law.

MYTH FOUR: Darwinism contains all fact and no faith. Again, this will be proven wrong. Darwinian evolution is a totally unproven philosophy that is based on highly speculative and hypothetical fossil evidence. Many supporters of Darwinism, including Charles Darwin himself, openly admit that physical evidence was, and still is, missing:

Darwin, Charles., *"Origin of Species"*, 6th edition, 1872, London, p. 413. (*2 quotes*)
"The number of intermediate varieties (missing links), which have formerly existed on the earth, (must) be truly enormous... Why then is not every geological formation and every stratum full of such intermediate links? Geology assuredly does not reveal any such finely graduated organic chain: and this, perhaps, is the most obvious and gravest objection which can be urged against my theory."

Easterbrook, Gregg., "Where Did Life Come From?," *Wired Magazine*, February, 2007, p. 108.
"What creates life out of the inanimate compounds that make up living things? No one knows. How were the first organisms assembled? Nature hasn't given us the slightest hint. If anything, the mystery has deepened over time."

Mayr, Ernst., Naturalist/Atheist, *"What is Evolution"*, Basic Books Publishing, 2001, p. 189.
"Wherever we look at the living biota, discontinuities are overwhelmingly frequent. The discontinuities are even more striking in the fossil record. New species usually appear in the fossil record suddenly, not connected with their ancestors by a series of intermediates."

Pagel, Mark., "Happy Accidents?", *Nature*, Vol. 397, February 25, 1999, p. 665.
"Paleobiologists flocked to these scientific visions of a world in a constant state of flux and admixture. But instead of finding the slow, smooth and progressive changes Lyell and Darwin had expected, they saw in the fossil records rapid bursts of change, new species appearing seemingly out of nowhere and then remaining unchanged for millions of years... patterns hauntingly reminiscent of creation."

Berlinski, David., *"The Deniable Darwin"*, Commentary, vol. 101, June 1996, p. 28.
"Unable to say what evolution has accomplished, biologists now find themselves unable to say whether evolution has accomplished it. This leaves the evolutionary theory in the doubly damned position of having compromised the concepts needed to make sense of life- complexity, adaptation, design- while simultaneously conceding that the theory does little to explain them."

Smith, Wolfgang., *"Teilhardism & the New Religion"*, Rockford, IL, Tan Books, 1988, (2 quotes), pp. 2, 5.
"We are told dogmatically that evolution is an established fact; but we are never told who has established it, and by what means. We are told, often enough, that the doctrine is founded upon evidence, and that indeed this evidence 'is henceforward above all verification, as well as being immune from any subsequent contradiction by experience; but we are left entirely in the dark on the crucial question wherein, precisely, this evidence consists. ... and yet the fact remains that there exists to this day not a shred of bona fide scientific evidence in support of the thesis that macroevolution transformations have ever occurred."

Lovtrup, Soren., Biologist, Former Darwinist/Naturalist, *"Darwinism: The Refutation of a Myth"*, New York: Croom Helm Publishers, 1987, p. 469.
"I believe that one day the Darwinian myth will be ranked the greatest deceit in the history of science'."

Rifkin, Jeremy., *Algeny*, New York; Viking Press, 1983, pp. 125, 134.
"What the record 'shows' is nearly a century of fudging and finagling by scientists attempting to force various fossil morsels and fragments to conform with Darwin's notions, all to no avail. Today the millions of fossils stand as very visible, ever-present reminders of the paltriness of the arguments and the overall shabbiness of the theory that marches under the banner of evolution."

Bethell, Tom., "Agnostic Evolutionists", *Harper's*, vol. 270, February 1985, p. 61.
"...it is also surely true that the positive evidence for evolution is very much weaker than most laymen imagine, and that many scientists want us to imagine. Perhaps as (Colin) Patterson says, that positive evidence is missing entirely."

Patterson, Colin., Senior Paleontologist at the British Museum of Natural History, speech at the American Museum of Natural History on: "Evolution and Creationism", New York, November 5, 1981, p. 2.
"Can you tell me anything you know about evolution, any one thing that is true? I tried that question on the geology staff at the Field Museum of Natural History and the only answer I got was silence!"

Muggeridge, Malcom., *"The End of Christendom"*, Grand Rapids: Eerdmans, 1980, p. 59.
"I myself am convinced that the theory of evolution, especially the extent to which it has been applied, will be one of the great jokes in the history books in the future. Posterity will marvel that so very flimsy and dubious a hypothesis could be accepted with the incredible credulity that it has."

Danson, R., *New Scientist*, 49:35, 1971. (2 quotes)

"The Theory of Evolution is no longer with us, because neo-Darwinism is now acknowledged as being unable to explain anything more than trivial change (micro-variation), and in default of some other theory, we have none... Despite the hostility of the witness provided by the fossil record, despite the innumerable difficulties, and despite the lack of even a credible theory, evolution survives. Can there be any other area of science, for instance, in which a concept as intellectually barren as embryonic recapitulation could be used as evidence for a theory?"

This book critically examines Darwinian science. I feel it would be a useful resource for teenagers, adults, and public school science teachers as a classroom supplement to challenge the unproven assumptions of Darwinism. But if it is used as a public school supplement please remember; **studying the scientific evidences and arguments against Darwinism is *NOT* teaching creationism!** Studying religious scriptures pertaining to 'the creation story' is teaching creationism. During my career as a science educator I have found it sadly amazing that most science teachers have studied the classic arguments *for* evolution, but have never read a serious document that outlines the scientific arguments *against* it. When educators do read both, they usually agree that the arguments against are significantly stronger.

Scattered throughout this book are quotes from experts in scientific fields relating to each evolution topic being discussed. Most of the more recent quotes I copied directly from the original source, but the others came from secondhand authors. [16] I am trusting that those quotes were accurately presented by them. If you find that any are misquoted, please contact me and I will revise them immediately. It is my goal to have accurate representations of each expert's words, because these experts are far more knowledgeable than I in each of their respective fields. Consider their words carefully. All of the scientific data presented here can be found in the reference books listed on page 96 unless an alternate source is cited. Some quotes are several decades old and are used to show the long term consistency of these Darwinian arguments. I have attempted to create an introductory document that defines terms related to Darwinism, and presents solid scientific arguments challenging its unproven assumptions. This book is a simplified compilation of work already in print. It takes complicated terminology and ideas, often intimidating to lay people, and reduces them to an understandable level. I hope this book helps students, parents, and teachers better understand Darwinism, especially the scientific arguments leveled against it.

David R. Browning

CHAPTER 2: WHAT EXACTLY IS EVOLUTION?

This book is designed to closely examine the term 'Darwinism', because the term 'evolution' is scientifically vague and can have multiple interpretations. Many scientists feel it is more accurate to sub-divide the word 'evolution' into these two specific terms:

I. MACROEVOLUTION ("Darwinism": massive genetic change... amoeba to man) This refers to slow and gradual genetic change from one original spontaneously generated cell, into all of the species that have ever lived on earth.

II. MICROEVOLUTION ("Micro-genetic variation": genetic variation within a gene pool) This refers to observable, measurable, but extremely small genetic variations that occur in every species.

Some scientists argue that macroevolution is simply millions of years of continuous microevolution, and that if there is scientific evidence for microevolution, then that automatically proves macroevolution. Opponents argue that microevolution is observable and proven, but no tangible evidence of macroevolution exists. Still others argue that the entire macroevolution topic has very little scientific merit, because Darwinism is technically outside the realm of modern operational science. Modern operational science always attempts to apply observable, repeatable, and scientific testing processes to verify a hypothesis, but all origin theories are pure speculation about the unobserved, untestable, and unrepeatable ancient earth past. If that is true, Darwinism would be relegated more to the realm of philosophy or religion, rather than science. Dr. Michael Denton and many other scientists articulate this argument well:

"The overriding supremacy of the myth (Darwinism) has created a widespread illusion that the theory of macroevolution was all but proved 100 years ago and that all subsequent biological research- paleontological zoological and in the newer branches of genetics and molecular biology- has provided ever-increasing evidence for Darwinian ideas. Nothing could be further from the truth. The fact is that the evidence was so patchy 100 years ago that even Darwin himself had increasing doubts as to the validity of his views, and the only aspect of his theory which has received any support over the past century is where it applies to microevolutionary phenomena. His general theory, that all life on earth had originated and evolved by a gradual successive accumulation of fortuitous mutations, is still, as it was in Darwin's time, a highly speculative hypothesis entirely without direct factual support and very far from that self-evident axiom some of its more aggressive advocates would have us believe." [8]

Simons, Andrew M., "The Continuity of Microevolution and Macroevolution," *Journal of Evolutionary Biology,* **15, 2002, pp. 688-701.**
   "A persistent debate in evolutionary biology is one over the continuity of microevolution and macroevolution -- whether macroevolutionary trends are governed by the principles of microevolution."

Stern, David L., "Perspective: Evolutionary Developmental Biology and the Problem of Variation," *Evolution,* **54, 2000, pp. 1079-1091. (2 quotes)**
   "One of the oldest problems in evolutionary biology remains largely unsolved... Historically, the neo-Darwinian synthesizers stressed the predominance of micro-mutations in evolution, whereas others noted the similarities between some dramatic mutations and evolutionary transitions to argue for macromutationism."

Prothero, Donald R. Ph.D., "**Punctuated Equilibrium at Twenty: A Paleontological Perspective**", *Skeptic*, vol. 1, no. 3, Fall 1992, pp. 38-47.

> *"If species sorting is real, then the processes operating on the species level, (macroevolutionary processes), are not necessarily the same as those operating on the level of individuals and populations, (microevolutionary processes). In other words, macroevolution may not just be microevolution scaled up."*

Kautz, Darrel., *"The Origin of Living Things"*, 1988, p. 6.

> *"People are misled into believing that since microevolution is a reality, that therefore macroevolution is such a reality also. This is sheer illusion; for there is no scientific evidence whatever to support the occurrence of biological change on such a grand scale."*

Denton, Michael., *"Evolution: A Theory in Crisis"*, **Adler & Adler, Bethesda, Md, 1986, p. 56.**

> *"There is no doubt that as far as his macroevolutionary claims were concerned, Darwin's central problem in 'The Origin of Species', lay in the fact that he had absolutely no direct empirical evidence in the existence of clear-cut intermediates, that evolution on a major scale had ever occurred, and that any of the major divisions of nature had been crossed gradually through a sequence of transitional forms."*

Lewin, Roger., biochemist, former editor of New Scientist and science writer, "Evolutionary- Theory Under Fire: An Historic Conference in Chicago Challenges the Four-Decade Long Dominance of the Modern Synthesis," *Science*, **Vol. 210, 21 November 1980, pp. 883-887.**

> *"The changes within a population have been termed microevolution, and they can indeed be accepted as a consequence of shifting gene frequencies. Changes above the species level-involving the origin of new species and the establishment of higher taxonomic patterns- are known as macroevolution. The central question of the Chicago conference was whether the mechanisms underlying microevolution can be extrapolated to explain the phenomena of macroevolution. At the risk of doing violence to the positions of some of the people at the meeting, the answer can be given as a clear... No."*

A portion of the material in this book will consider a fascinating scientific question: is it possible that there are intelligent life forms existing elsewhere in the universe, more advanced than humans, which may have played some part in engineering life on earth? Modern science has proven *(via the thermodynamic and entropy laws, spontaneous generation, biogenesis, and cell theory)* that matter and life could not have originated by any random, natural mechanisms, and yet... matter and life exist! If all ***natural*** creative causes have been ruled out, that only leaves ***super***-natural creative causes. Atomic and cellular complexity, as well as the seemingly miraculous informational coding seen in DNA, continues to convince scientists around the world that matter, energy, and life must have had an external, intelligent, causal agent. Some people refer to this agent as 'God', while others are suggesting it was an advanced extraterrestrial. For the sake of discussion in this book, this hypothetical being will be called a 'designer'. Evolutionary biologist and celebrated atheist author, Richard Dawkins, has spent his entire career writing, teaching, and debating against intelligent design, and yet the reality of today's complexity evidence recently led him to admit the following:

(Atheist/Naturalist Richard Dawkins, Photo: Oxford University)

"It (Intelligent design) could come about in the following way. It could be that at some earlier time, somewhere in the universe, a civilization evolved by some type of Darwinian means to a very, very high level of technology, and designed a form of life that they seeded onto perhaps, this planet. Now, that is a possibility, and an intriguing possibility. And I suppose it's possible that you might find evidence for that, if you look at the details of biochemistry and molecular biology. You might find a signature, of some sort of designer. And that designer could well be a higher intelligence from elsewhere in the universe, but it (that original intelligence) couldn't have first jumped into existence spontaneously." [9]

So, even naturalist Richard Dawkins concedes that life is incapable of spontaneously generating and that design evidences have forced him to conclude that earth life was probably engineered by some type of higher intelligence. But, if earth life is known to be too complex to spontaneously arise, how could an extraterrestrial designer of even *higher* complexity spontaneously arise first? His dilemma is obvious; his reasoning is self-contradictory on many levels.

These intelligent design arguments are becoming so evident, that scientists from every continent have begun signing a document called: **"A Scientific Dissent from Darwinism"**. Only the world's top research scientists holding PhD and MD degrees are permitted to sign this document. As of March 2009, over 800 signatures have been registered. These researchers represent some of the greatest scientific minds currently in academia *(www.discovery.org, 2009)*. It is also interesting to read statements from contemporary **Nobel Award-Winning Scientists** pertaining to design. These brilliant researchers have a thorough understanding of the evidences for, and against, Darwinism. Look at this partial listing of Nobel names who are confirmed intelligent design apologists: Einstein, Planck, Schroedinger, Heisenberg, Millikan, Townes, Schawlow, Phillips, Bragg, Marconi, Compton, Penzias, Mott, Rabi, Salam, Hewish, Taylor, Carrel, Eccles, Murray, Chain, Wald, Ross, Barton, Kohn, and Smalley *(www.nobelist.tripod.com, 2009.)* I find it curiously interesting that many naturalists ridicule fellow scientists who agree with the intelligent design insights of Albert Einstein and Max Planck, while admitting that they are two of the world's most brilliant and creative minds in the fields of mathematics, chemistry, and atomic physics:

Einstein, Albert., various quotes as cited by Brian Denis in: *"Einstein: A Life"*, New York, John Wiley and Sons, 1996, pp. 119, 186.

"The deeper one penetrates into nature's secrets, the greater becomes one's respect for God. We see a Universe marvelously arranged and obeying certain laws, but only dimly understand these laws. Our limited minds cannot grasp the mysterious force that moves the constellations. ... Science without religion is lame, religion without science is blind. ... The more I study science, the more I believe in God. ... I want to know God's

thoughts, the rest are details. That, it seems to me, is the attitude of even the most intelligent human being toward God."

Planck, Max., as cited by Kurt Eggenstein in: *"Materialistic Science on the Wrong Track"*, **1984.**
"As a physicist, that is, a man who had devoted his whole life to a wholly prosaic science, the exploration of matter, no one would surely suspect me of being a fantast. And so, having studied the atom, I am telling you that there is no matter as such! All matter arises and persists only due to a force that causes the atomic particles to vibrate, holding them together in the tiniest of solar systems, the atom. Yet in the whole of the universe there is no force that is either intelligent or eternal, and we must therefore assume that behind this force there is a conscious, intelligent Mind or Spirit. This is the very origin of all matter."

The "design" opinions of Planck, Einstein, and other Nobel scientists are relevant, interesting, and legal to study in public schools. But when discussing Darwinism in classrooms, teachers would be wise to refrain from assigning homework or administering tests, since no origin models have been empirically proven, and probably never will be. As with all classroom discussions, students must be allowed their 1st Amendment rights of free speech concerning their religious and scientific opinions. All student viewpoints concerning origins should be allowed and respected in public schools, and any curriculum based solely on one evolutionary theory, or on religious scripture, should be limited to private schools. The open-forum approach to evolution is consistent with the Constitution, Civil Rights, religious neutralism, scientific objectivity, educational effectiveness, academic freedom, and general fairness.

The next two pages give detailed definitions and evidences for both macro and micro evolution, and the remainder of the book will examine the 10 iconic evidence categories taught by classic Darwinism. This book investigates the current data that claims to support these iconic evidences. Read, study, think, and decide for yourself if Darwinian evolution seems reasonable to you based on scientific evidence, and beware of scientists on either side of the discussion who blindly reject solid data due to their personal biases.

MACROEVOLUTION: (aka: "DARWINISM, NATURALISM")

THEORIZES: That the matter that makes up the universe is assumed to be eternal *(because matter can not be created by any natural process)*, self creating, self existing, self directing, and that it self-assembled from disorder to order by random chance. It assumes that DNA and the first living cell both self-assembled by spontaneous generation, that the DNA code positively mutates *(simultaneously in males and females)* and reproduces into higher and more complex organisms through millions of years of gradual genetic mutation *(birth defects)*, that macroevolution is too slow to be directly observed and proven but is continuously ongoing, that all earth organisms originated from that first 'primitive' cell, that the gene pool is increasing in size and complexity, that all life is connected and can be traced by fossil evidence, and that the earth is 4.6 billion years old *(based on today's rate of geologic sedimentation)* with life originating at the Cambrian layer.

PROBLEMS WITH THE THEORY: Macroevolution contradicts the laws of physics, chemistry and energy *(conservation of mass, laws of thermodynamics, laws of energy and decay, laws of classification and order, laws of cause and effect, laws of motion...)*, because these systems all operate in a perpetual state of decay, or status quo at best. No known systems randomly go from disorder to order without intelligent input. It is generally agreed that the universe did have a definite beginning *(big bang)* and will have an ending *(heat death)*. Spontaneous generation data, the modern biological cell theory, and mathematical probability *denies* the possibility of a living cell's self-assembly and functionality. Contemporary scientists know living cells come only from other living cells, and they cannot assemble a living cell from its basic elemental building blocks. The gene pool is breaking down over time, not increasing in complexity. No vertical macroevolution is currently being observed or has ever been found in the fossil record, only microvariation is proven to exist. Mutations are not a plausible vehicle for vertical change because mutations are random, rare, and usually destructive. There is no fossil evidence to connect different groups of organisms *(like amphibians and reptiles)* only indirect and disputable evidence. The Cambrian Explosion shows that most major life groups started virtually simultaneously, and there is no fossil 'chain of descent' from a common ancestor. Macroevolution demands incredibly long time periods to even be remotely plausible. The 4.6 billion-year-old Darwinian earth date is unscientifically based on the rate of today's sediment buildup. There is no proof for, and there is much proof against, the earth's sedimentation rate remaining constant throughout all of the earth history *(due to proven catastrophes like super-volcanoes, mega-floods, and meteor strikes...)*. DNA and instincts are two types of biological information that are virtually unexplainable via Darwinism.

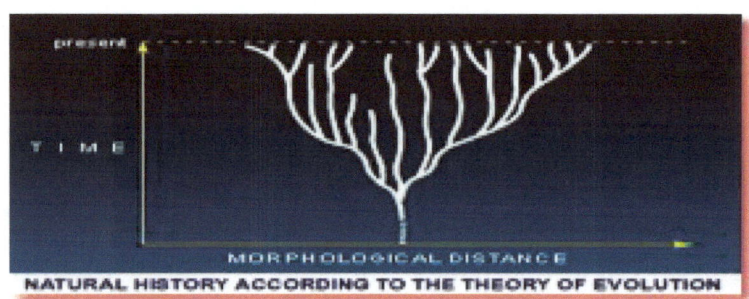

(2009, by H. Yahya)

The classic Darwinian "tree of descent" from a common ancestor.

MICROEVOLUTION: (aka: "HORIZONTAL EVOLUTION, MICROVARIATION")

THEORIZES: That matter and energy can not create itself *(laws of thermodynamics, entropy)*, that time, space, matter and energy had a specific beginning *(Big Bang)* and will have a specific ending *(universal heat death)*, that all matter and systems are in a state of decay, that complex systems can not go from disorder to order by random chance, that most modern taxonomical groups appear simultaneously in multiple, abrupt beginnings, in easily classified groups starting near the Cambrian layer *(including vertebrates)*, that organisms throughout all geologic sediment layers show no proven macroevolutionary fossil linkage with other taxonomical groups *(no descent from common ancestor)*, and that only microvariational change is observed and proven within groups *(like the dog family, peppered moths, finch beaks, antibiotic resistance, etc...)*. It recognizes that gene pools are showing no increase in complexity, but they are diversifying and decaying over time. It recognizes that genetic mutations are nearly always destructive and not constructive, that matter and life easily exceed accepted scientific thresholds for external design *(which suggests a designer)*, and that extinction, mutation, and defects are to be expected over time, not increasing complexity. It recognizes that cells are far too complicated to spontaneously generate from inanimate elemental matter *(modern cell theory and biogenesis agree)* which proves that living cells can only come from similar living cells. Microevolution is unconcerned with earth age since it is irrelevant, but agrees that the classic geologic column is a poor measure of earth age due to the lengthy list of proven catastrophic events that have altered sedimentation and radiometric decay rates. Additional types of earth dating data, often considered more scientific, measurable, and reliable, should be equally considered when estimating hypothetical earth ages.

PROBLEMS WITH THE THEORY: All of the above concepts are currently in agreement with today's known scientific data, however, some scientists disregard even the *possibility* of an extraterrestrial designer on the basis that it has not yet been directly observed, studied, or tested in a closed and controlled laboratory setting.

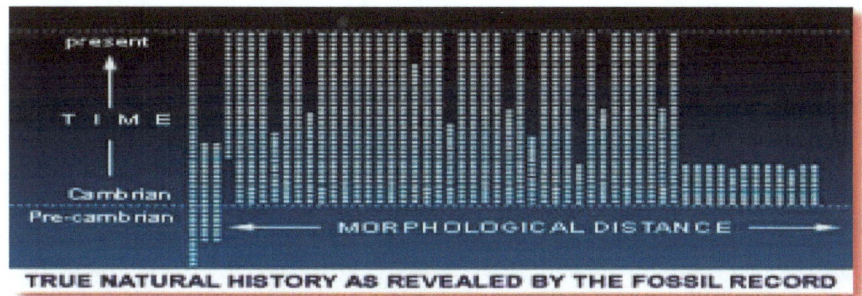

(2009 by H. Yahya)

The "Biological Big Bang", the simultaneous Cambrian appearance of multiple life groups.

CHAPTER 3:   IS DARWINISM A THEORY OR A RELIGION?

Webster defines religion as: *"A belief in a particular system of faith; belief without proof in a doctrine or philosophy, an unproven dogma"*.

Science, on the other hand, always seeks observable and measurable data to answer specific questions about matter and life processes. Darwinian evolution, as a true science, lives or dies with its ability to prove the following ten concepts:

   1. Matter's self-creation,
   2. Earth age,
   3. Spontaneous cell formation,
   4. Genetic increase via mutation,
   5. Transitional fossil chains,
   6. Natural selection causing speciation,
   7. Homology,
   8. Embryology,
   9. Vestigial organs, and
   10. Punctuated equilibrium.

These ten iconic 'proofs' are always presented by Darwinists as evidence that their theory is scientifically sound. If no hard evidence exists to support these concepts, then Darwinism is rendered a philosophy or a religion, rather than a fact-based field of science. As you are about to see, many of today's top research scientists argue that all ten of these Darwinian concepts are void of evidential support, thus relegating the theory to the realm of a religion. Many top scientists have been arguing since the 1960's that Darwinism has reached an intellectual and evidential dead end, and that current technological breakthroughs are only verifying this reality. For instance:

Ruse, Michael., "Saving Darwinism from the Darwinists", *National Post*, May 13, 2000, B3.
   "...Evolution is a religion. This was true of evolution in the beginning, and it is true of evolution still today."

Hsu, K. J., *"Journal of Sedimentary Petrology"*, 56(5), 1986, pp. 729-730. (several quotes)
   "We have all heard of (Darwin's book) The Origin of Species, although few of us have had time to read it... A casual perusal of the classic made me understand the rage of Paul Feyerabend.... I agree with him that Darwinism contains "wicked lies"; it is not a "natural law" formulated on the basis of factual evidence, but a dogma, reflecting the dominating social philosophy of the last century."

Cartmill, Matt., "Four Legs Good, Two Legs Bad", *Natural History*, vol. 92, 1983, p. 77.
   "A myth, says my dictionary, is a real or fictional story that embodies the cultural ideals of a people or expresses deep, commonly felt emotions. By this definition, myths are generally good things- and the origin (evolution) stories that paleontologists tell are necessarily myths."

Thomsen, Dietrick E., "A Knowing Universe Seeking to be Known", *Science News*, vol. 123, February 1983, p. 124.

"Objections will be raised by those who want to insist that the material universe is all the reality there is. Materialism (Naturalism) of this kind is a doctrine that anyone may choose as a working hypothesis or as a religion, but I am unaware of a proof of it."

Lipson, H. S., "A Physicist Looks at Evolution", *Physics Bulletin*, vol. 31, 1980, p. 138.   (Several quotes)

"In fact, evolution became in a sense a scientific religion; almost all scientists have accepted it and many are prepared to 'bend' their observations to fit in with it. To my mind, the theory does not stand up at all. If living matter is not, then, caused by the interplay of atoms, natural forces, and radiation, how has it come into being? ...I know that this is anathema to physicists, as indeed it is to me, but we must not reject a theory that we do not like if the experimental evidence supports it. ... I have always been slightly suspicious of the theory of evolution because of its ability to account for any property of living beings."

Wysong, R.L., *"The Creation-Evolution Controversy"*, Inquiry Publisher, 1981, p. 455.

"Evolution requires plenty of faith; a faith in L-proteins that defy chance formation; a faith in the formation of DNA codes which, if generated spontaneously, would spell only pandemonium; a faith in a primitive environment that, in reality, would fiendishly devour any chemical precursors to life; a faith in experiments that prove nothing but the need for intelligence in the beginning; a faith in a primitive ocean that would not thicken, but would only haplessly dilute chemicals; a faith in natural laws of thermodynamics and biogenesis that actually deny the possibility for the spontaneous generation of life; a faith in future scientific revelations that, when realized, always seem to present more dilemmas to the evolutionists; faith in improbabilities that treasonously tell two stories – one denying evolution, the other confirming the Creator; faith in transformations that remain fixed; faith in mutations and natural selection that add to a double negative for evolution; faith in fossils that embarrassingly show fixity through time, regular absence of transitional forms and striking testimony to a worldwide water deluge; a faith in time which proves to only promote degradation in the absence of mind; and faith in reductionism that ends up reducing the materialist's arguments to zero and `forcing the need to invoke a supernatural Creator."

Azar, Larry., "Biologists, Help!", *Bioscience*, vol. 28, November 1978, p.714.

"If a biologist teaches that between two existing fossils there was a non-existing third, and perhaps several others, is he not really like the man of religious faith who says: 'I believe, even though there is no evidence?'"

Patterson, Colin., Paleontologist, British Museum of Natural History, London, *"Evolution"*, 1977, p. 150

"(Karl) Popper warns of a danger: a theory, even a scientific theory, may become an intellectual fashion, a substitute for religion, an entrenched dogma. This has certainly been true of evolutionary theory."

Sokolov, B.S., 'The Current Problems of Paleontology and Some Aspects of Its Future", *Paleontological Journal*, vol. 9, no. 2, 1975, p. 137.

"I know geologists who regard the whole of Darwin's theory and the present-day synthetic theory of evolution, which do in fact have weak spots, as a type of religion, but we may readily imagine the chaos that would face us in geology were the evolutionary concept to become a myth."

Rosazak, T., *"The Unfinished Animal"*, 1975, pp. 101-102.

"The irony is devastating. The main purpose of Darwinism was to drive every last trace of an incredible God from biology. But the theory replaces God with an even more incredible deity--- omnipotent chance."

Matthews, Harrison L., Naturalist/Darwinist, *"Introduction to the Origin of Species"*, London: Dent and Sons, 1971, p. x.

"The fact of evolution is the backbone of biology, and biology is thus in the peculiar position of being a science founded on an unproved theory- is it then a science or a faith? Belief in the theory of evolution is thus exactly parallel to belief in special creation- both are concepts which believers know to be true... but neither, up to the present, has been capable of proof."

COULD DARWINIAN 'THEOLOGY' BE DANGEROUS?

Darwinian evolution insists that matter and life spontaneously generates, and that life gradually and steadily increases in complexity via mutation and natural selection or, **"SURVIVAL OF THE FITTEST"**. This iconic evolutionary process claims to create new organs and organisms, eliminates the weakest life forms, and propagates the genetics of the strongest and most able. This is the engine that Darwinists claim drives random disorder toward ordered complexity. Evolutionists insist that their theory poses no danger to society, and that it is simply an intellectual and scientific field of study. But consider this: if Darwinian evolution is the force that creates, modifies, and increases the complexity of life on earth, then several truths are inescapable:

1. All predators should attack and kill other life forms because it is helpful to speciation and evolution. Predators weed out the sick, old, and helpless, leaving only the strongest and smartest.

2. Evolutionary laws affect and help all life forms equally and should be *encouraged* whenever possible.

3. Humans are just another organism that holds no higher intrinsic value than a plant, insect, or bacterium. We are simply a higher, and more recently evolved carbon-based life form, that will some day be surpassed by an even more highly evolved animal.

4. Arbitrary man-made laws that prevent humans from killing would be counter-productive to the forces of evolution because they oppose survival of the fittest. For example: if person A is capable of killing person B and taking all of his possessions for himself and his family, person A has eliminated a weaker organism and is better able to survive and pass on his genetics to the next generation. Likewise, person A should have sexual intercourse with as many humans as possible to give his/her superior genetics the best possible chance to be propagated to the next generation. Most human laws and morals would be rendered meaningless, because most run contrary to evolutionary processes.

5. Marauding "street gangs" should not only be allowed, but encouraged, because they perfectly exemplify survival of the fittest within the human population.

6. Similarly, wars between nations beautifully exemplify survival of the fittest, and would improve evolutionary progress as *'stronger'* nations plunder and eliminate *'weaker'* nations. Ethnic cleansing was/is simply Darwinism in action.

7. Hitler's attempt to genetically improve his nation by destroying weaker races/nations was driven by evolutionary 'eugenics' motives and should be applauded, encouraged, and repeated.

8. Abortion and euthanasia are in absolute agreement with evolutionary laws, as long as the human population is stable and does not decline dangerously. Eliminating excess and under-productive humans would allow more resources to go to the strongest.

9. All birth defected life forms, especially humans, should be exterminated ASAP to improve, protect, and positively evolve their gene pools.

10. Non-productive elderly humans would be fair game for extermination and possession removal so that younger and better fit humans would benefit.

11. Most world religions would be rendered completely meaningless, because their 'laws' totally contradict evolutionary law. Most religions are based on the precept that a being of very high intellectual sophistication designed man for a purpose, and established for him specific laws of social behavior, most of which run contrary to evolutionary logic.

12. Some contemporary scientists and theologians suggest that a designer might have used evolution as its life creating mechanism. But that begs one question: what intelligent creator would use totally random processes to bring about purposeful, directed, specified, and guided creative goals, when evolutionary processes demand the diametric opposite? The contrast is obvious. To me, logic dictates that one is true, the other is false. Purposeful creation and random-chance evolution cannot co-exist.

Now, those Darwinian observations make perfect sense within a purely evolutionary world-view, but who would want to live in a world governed by those laws? Probably only murderers, rapists, thieves, gang lords and dictators. Most humans prefer the opposite of what Darwinian evolution demands. Most people want laws that protect the weak, and punish human predators. Macroevolution celebrates and encourages predation.

CHAPTER 4: ARE THERE ANY DESIGN EVIDENCES?

1. *"When in the Course of human events, it becomes necessary for one people to dissolve the political bands which have connected them with another, and to assume among the powers of the earth the separate and equal station to which the Laws of Nature and of Nature's **God** entitles them..."*
2. *"We hold theses truths to be self-evident, that all men are **created** equal. That they are endowed by their **Creator** with certain unalienable rights; that among these are life, liberty and the pursuit of happiness..."*
3. *"We, Therefore, the Representatives of the United States of America, in General Congress, Assembled, appealing to the **Supreme Judge** of the world for the rectitude..."*
4. *"And for the support of this Declaration, with a firm reliance on the protection of **Divine** Providence, we mutually pledge to each other our lives, our fortunes..."*
5. (The National Motto of the United States of America)..............*"In **God** We Trust"*
6. (The United States Pledge of Allegiance)......*"...one nation, **Under God**, indivisible..."*

The United States of America is a unique experiment in governmental and educational history. It was founded under the assumption that a divine Creator not only exists, but formed mankind with specific purposes and rights. If true, the scientific method of inquiry should find evidence to verify the presence of this Creator. Since the passage of the Declaration of Independence, the Pledge of Allegiance, and the National Motto, science has neither proven, nor disproven, a creator. Public school science curricula may legally discuss the generic topic of creator, as long as no specific religion is promoted or denied, and educators must remain absolutely neutral on the topic of religion. *(Students, on the other hand, have no such limits on their religious free speech.)* The topic of creator easily fits within the realm of scientific discussion as long as the curriculum is examining scientific evidences and not religious scripture. Some educators contend that science is incapable of studying entities that are invisible, but science quite often proves the existence of matter and energies that have never been seen, like: atoms, gravity, quarks, air, and radio waves. Nobody questions their existence. Those entities have been proven to exist because of overwhelming *indirect* evidence, not visual evidence. Many of the world's most respected researchers have also presented solid scientific arguments that a higher creative intelligence exists, and played a part in forming time, space, matter, energy, and organic life. A partial list of these scientists includes:

Physics> Newton, Faraday, DaVinci, Copernicus, Maxwell, Fleming, Von Braun, Morse...
Chemistry> Boyle, Dalton, Kelvin, Ramsay, Woodward ...
Biology> Ray, Linnaeus, Mendel, Agassiz, Hooke, Carver, Fabre...
Astronomy> Copernicus, Galileo, Kepler, Herschel, Maunder...
Mathematics> Pascal, Leibnitz, Euler, Planck, Einstein ...
Geology> Steno, Brewster, Buckland, Cuvier...
Medicine> Virchow, Lister, Pasteur...

That list reads like a 'Who's Who' of the most brilliant minds in scientific history. There are also thousands of contemporary scientists who currently see scientific evidences for intelligent design, like: Dr. Dean Kenyon, Dr. A. E. Wilder-Smith, Dr. W. R. Thompson, Dr. Melvin A. Cook, Dr. Henry M. Morris, Dr. Walter

Lammerts, Dr. Frank Marsh, Dr J. J. Duyvene De Wit, Dr. Thomas G. Barnes, Dr. Dmitri Kouznetsov, Professor Leonid Korochkin, Sir Fred Hoyle, Dr. Chandra Wickramasinghe, the 800+ top research scientists that have signed the *"Scientific Dissent From Darwinism"* document *(www.discovery.org)* as well as the Nobel scientists mentioned earlier *(www.nobelist.tripod.com)*. Their scientific arguments for design are legal and interesting to study in public school science classrooms.

Every analysis of Darwinism should begin by taking a long hard look at the **'Big Bang Theory'**, because one of the strongest arguments for intelligent design can be made by examining the origin of matter and energy. Nearly a century ago, scientists successfully proved that matter and energy can *not* be created or destroyed by any natural mechanism. These cardinal truths are known as the laws of conservation of mass and energy, and the laws of entropy. These fundamental laws have been repeatedly verified, are universally accepted as scientific truth by all scientists, and they appear in chapter one of every modern science text book. Because it has been proven that no natural mechanism can create matter, ***and yet matter exists,*** naturalists logically argued that matter must have always been in existence. If it is true that matter has always existed, then:

1. matter and energy must be eternal and had no beginning moment in time,
2. matter and energy must be self-sustaining,
3. matter and energy must be self-directing solely by random and natural processes.

But, several decades ago, shocking evidence was discovered which began to refute these classic naturalistic assumptions. Astronomers were seeing evidence strongly suggesting that there had been a singular, explosive, and creative moment for the universe... a *'Big Bang'*. Naturalists initially resisted these new findings because the evidence suggested that matter and energy had a creative moment in time, which totally contradicted their above assumptions. The new evidence also closely matched 'creationist' scriptures, which they openly ridiculed. Once documentation for the Big Bang became overwhelmingly accepted, naturalists had to restructure their core beliefs. Why? Because it is now generally agreed that time, space, matter, and energy:

1. Had a singular creative moment and is steadily expanding outward: *(The Big Bang)*.
2. Will have an 'ending' moment: *(Heat Death, 2nd law of Entropy)*.
3. Cannot be created or destroyed, and yet it exists, *(Conservation Laws)*,
4. Is unwinding from that singular creative moment: *(Entropy Laws)*.
5. Will forever expand because it lacks sufficient mass to re-collapse, or, 'recycle'.

Every fact that had to be true for the universe to form by natural cause has been proven false by modern science. Therefore, naturalists have to accept one of two realities. Either; time, space, matter and energy were formed by a **SUPER**-natural cause, which requires that some intelligent creative agent(s) exists outside of the natural laws of science, or, all of the hallowed and revered scientific laws governing the conservation of mass, energy, and entropy, are **WRONG**. That would throw all fields of science into total chaos. Naturalists, of course, refuse to accept either choice, but one choice has to be correct, and the other one incorrect. Listen to contemporary scientists speak in confused amazement about the creative moment of the universe:

Easterbrook, Gregg., "The New Convergence: Science + Religion", *Wired Magazine*, December 2002, p. 49.

"In recent years, Allan Sandage, one of the world's leading astronomers, has declared that the Big Bang can be understood only as a 'miracle.' Charles Townes, a Nobel-winning physicist and co-inventor of the laser, has said that discoveries of physics 'seem to reflect intelligence at work in natural law'."

Kaku, Michio., quoted Nina L. Diamond, *Voices of Truth*, 2000, pp. 333-334.   *(2 quotes)*

"The strength and weakness of physicists is that we believe in what we can measure. And if we can't measure it, then we say it probably doesn't exist. And that closes us off to an enormous amount of phenomena that we may not be able to measure because they only happened once. For example, the Big Bang. ... That's one reason why they scoffed at higher dimensions (God) for so many years. Now we realize that there's no alternative."

Darling, D., "On Creating Something From Nothing", *New Scientist*, v. 151, September 1996, p. 49. *(2 quotes)*

"...the biggest deal of all- is how you get something out of nothing. ...Either there is nothing to begin with, in which case there is no quantum vacuum, no pregeometric dust, no time in which anything can happen, no physical laws that can effect a change from nothingness into somethingness; or there is something, in which case that needs explaining."

Sandage, Alan., (winner of the Crawford Prize in astronomy), and Willford, J.N. *"Sizing Up The Cosmos: An Astronomers Quest."*, New York Times, March 12, 1991, p. B9.

"I find it quite improbable that such order came out of chaos. There has to be some organizing principle. God to me is a mystery, but is the explanation for the miracle of existence... why there is something, instead of nothing."

Davies, Paul., *"The Edge of Infinity"*, New York: Simon and Schuster, 1981, p. 161.

"... (The Big Bang) represents the instantaneous suspension of physical laws, the sudden, abrupt flash of lawlessness that allowed something to come out of nothing. It represents a true miracle---transcending physical principles...."

Rifkin, Jeremy., *"Entropy: A New World View"*, New York: Viking Press, 1980, p. 55. *(2 quotes)*

"The Entropy Law says that evolution dissipates the overall available energy for life on this planet. Our concept of evolution is the exact opposite. We believe that evolution somehow magically creates greater overall value and order on earth. Now that the environment we live in is becoming so dissipated and disordered that it is apparent to the naked eye, we are for the first time beginning to have second thoughts about our views on evolution, progress, and the creation of things of material value. ... Evolution means the creation of larger and larger islands of order at the expense of ever greater seas of disorder in the world. There is not a single biologist or physicist who can deny this central truth. Yet, who is willing to stand up in a classroom or before a public forum and admit it?"

Davies, Paul., *"Universe in Reverse"*, Second Look, London: King's College, Sept.'79, p. 27.

"The greatest puzzle is where all the order in the universe came from originally. How did the cosmos get wound up, if the second law of thermodynamics predicts asymmetric unwinding towards disorder?"

A strong scientific case can be made, and *has* been made, that the creation of time, space, matter and energy had to be controlled by a causal agent(s) existing outside the laws of nature. If true, there should be a trail of material evidence for scientists to find that indicates the presence of this designer. Many scientists now believe that the evidential fingerprint trail leads directly to the front door of organic life.

INTELLIGENT DESIGN: ONE BIG FIGHTING TERM

Just mentioning the term **'intelligent design'** sends many naturalists into an anger-filled tirade, but can rational, professional, and intelligent scientists make a case for the presence of organic intelligent design without invoking a holy war? Yes. Archeologists do it everyday in their field of study. As atheist Richard Dawkins pointed out several pages earlier, intelligent agents always leave markers, or, signatures that can be examined and studied like fingerprints. As you are about to see, increasing numbers of scientists are claiming that they have located and identified dozens of design signatures.

Meyer, S.C., "The Creation Hypothesis", Intervarsity Press, 1994, pp. 98, 102. *(2 quotes)*
> *"We have not yet encountered any good reason to exclude design from science. Design seems just as scientific, or unscientific, as its (Darwinian) competitors… Does a strictly materialistic evolutionary scenario or one involving intelligent agency or some other theory best explain the origin of biological complexity, given all relevant evidence?"*

Do the topics of **complexity, design,** and **information** fit within the classic scientific method of inquiry? Absolutely, they are *very* integral components of all fields of science. Whenever a new object is discovered and is being evaluated for origin, scientists make note of the number of variables that the item exhibits, and they then calculate the odds that random forces might have generated it. For example, archeologists routinely unearth ancient stone objects. They usually have no trouble determining if the objects were produced by human intelligence, or if they were simply carved by environmental erosion. Similarly, shapes etched onto ancient slabs of stone are examined in an attempt to detect the possible presence of an ancient language. Buildings, like the great pyramids, are studied and their complexity levels are compared to other known structures to determine the educational and technological capabilities of that civilization. Scientists are quite competent in their ability to discern between intelligently generated technologies, and naturally occurring earthen materials. Could rock faces, like those seen at Mt. Rushmore, be created on a stone cliff by random environmental forces?

(Internet Photo: www.nps.gov)

Yes, because erosion is capable of breaking apart rock. But what are the odds that mindless environmental erosion could cause complex facial features, perfectly resembling four famous American presidents? While *technically* possible, it would be highly unlikely. So mathematically unlikely in fact, that the notion of chance is quickly discarded and intelligent creativity would be logically assumed. Some probabilities far exceed the mathematical threshold of what is scientifically reasonable due to chance. These highly unlikely explanations would be discarded and the scientist would look for an alternative explanation that has more reasonable probabilities.

Now let's consider some organic scenarios. I am convinced that it is scientifically sound to argue that biological 'machines' are far too complex to form by chance, even if hundreds of millions of years are allowed. For example, consider the echolocation organs of dolphins and bats, both of which far exceed the functional sensitivity of man's most high-tech echolocation systems *(sonar and radar)*. Their complexity begs several questions. First, if man's *mechanical* echolocation devices could never form solely by random chance interaction of atoms, why should we assume that the even more complex *organic* echolocation systems of animals could form by random chance? Next, what about eyeballs, wings, livers and brains which have even higher states of complexity than echolocation organs? Thirdly, if we see no transitional fossils showing partially evolved organs, then doesn't that suggest that those organs must have appeared in a single mutational moment? And finally, if it is unreasonable to assume that a complex organ could form in a single naturalistic moment, how could a living cell, which has the physical complexity of a modern city, form by random chance processes? And remember, the more miniaturized a complex system is, the greater its design evidence. *(To be discussed further in chapter 4)*

A second important category of design evidence is that of pre-programmed **biological information,** and one example of this is **instincts**. Thousands of species of organisms exhibit incredibly complex, innate behaviors. True instincts, like the nest building of birds, are easily proven to be unlearned behaviors, and they represent vast quantities of stored information. The nest building instinct could never evolve slowly and gradually via Darwinian mechanisms, because whole and intact nests are necessary for the species' immediate survival and reproduction. Nest building represents complex data, it is stored in the bird's memory and genetics exactly like a detailed computer program, and it must have come into immediate existence in order to be useful. Naturalists have no plausible explanation for the origin of instinctual information, but ID proponents see instincts as intelligently designed data pre-programmed into the brains of organisms to enable them to survive. What quantity of information does an instinct represent? Industrial robotics engineers have been increasingly curious; some have even run experiments to see just how much coded instructional information has been stored in the brain of a bird in order to build a typical nest. This could be quantitatively measured by setting the raw materials needed to build a nest *(twigs, grass, mud, tree, etc...)* in front of an industrial robotic arm. I am confident that the computer program needed to build a crude bird nest would be impressively long, complex, and would far exceed chance formation probabilities.

West, Judy., "Birds Do It, Bees Do It… Now Robots?", *www.upenn.edu/pennnews/,* January, 2009. *(2 quotes)*
   "*The director of Penn Engineering's GRASP (General Robots Automation Sensing Perception) Lab, Vijay Kumar believes nature may hold the answers to some of the most complex challenges in the world of robotics. Specifically, he's interested in the way some animals exhibit collective behaviors—swarming, flocking (nest*

building) and the like—to accomplish a task, and how studying those behaviors could inform the design of large networked groups of robots. '...And remember, their nervous systems are known to be primitive. They don't have the capacity to do all the calculations for themselves, and they're not communicating with some central supervisor'."

Gitt, Werner., Director at the German Federal Institute of Physics and Technology, *"In the Beginning Was Information"*, Christliche Literatur-Verbreitung Publishing, 1997, p. 106.

"There is no known law of nature, no known process, and no known sequence of events which can cause information to originate, by itself, in matter."

Where did the original, innate, instinctual instructions come from, and how are they passed from generation to generation? Darwinists are baffled, but design proponents just smile and nod. Information is known to *only* come from intelligence. Do other forms of organic information exist, in addition to instincts? Yes. It is my opinion that the most convincing type of biological information known is **DNA,** the organic life code. It is as ordered and complex as any man-made informational code, and is exactly analogous to Morse code and computer programming code. All known information systems *(like books, maps, oral and written languages, Braille, blueprints, and military codes)* have the following characteristics in common:

1. They are all destroyed by random change.
2. They all have order, *(i.e. repeating patterns)*.
3. They all have specified complexity *(which is composed of order + structure)*.
4. The simultaneous presence of all three is always evidence of an intelligent origin.

Consider this: researchers for the **SETI** program *(the **S**earch for **E**xtra-**T**errestrial **I**ntelligence)* spend millions of dollars each year scanning the universe with huge radio-astronomy satellite dishes trying to detect the presence of intelligent extraterrestrial life. What type of signals are their huge receivers programmed to detect? They are searching for stray electromagnetic waves and/or light pulses that exhibit ***orderly repeating patterns,*** with ***specified complexity,*** which would be ***destroyed by random change.*** These are the same characteristics that are convincing scientists that DNA must be an intelligently engineered form of organic information, and could not have originated from random collisions of atoms. Organic chromosomes exhibit all of the classic scientific attributes of intelligently generated and coded information. Even world renowned atheists agree with this:

Dawkins, Richard., Naturalist, Zoologist and atheistic author, Professor for the Public Understanding of Science, Oxford University, *"River out of Eden: A Darwinian View of Life,"* Phoenix: London, 1996, pp. 19-20.

"After Watson and Crick (co-discoverers of the genetic code), we know that genes themselves, within their minute internal structure, are long strings of pure digital information. What is more, they are truly digital, in the full and strong sense of computers and compact disks, not in the weak sense of the nervous system. The genetic code is not a binary code as in computers, nor an eight-level code as in some telephone systems, but a quaternary code, with four symbols. The machine code of the genes is uncannily computer-like. Apart from differences in jargon, the pages of a molecular-biology journal might be interchanged with those of a computer engineering journal."

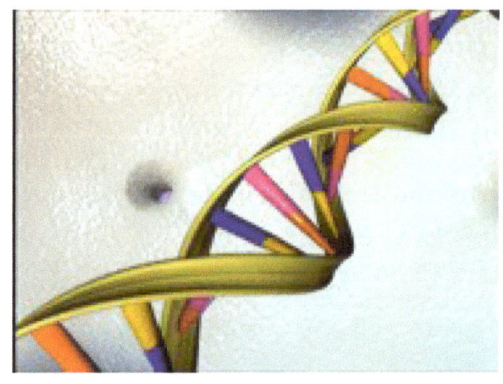
(Internet Photo from: www.genome.gov)

Examine any advanced genetics book and you will find these amazing DNA statistics:

1. Each human cell has 30,000 genes = 3.15 *BILLION* base pairs = one thousand books containing 500 pages of complex, ordered, coded information.

2. The amount of information that could be stored in a pinhead's volume of DNA is equivalent to a pile of books 500 times as tall as the distance from earth to moon.

3. Each human is capable of producing 10^{2017} genetic variations of sperm/egg.

4. Compare that to the number of atoms in the entire universe: 10^{80}.

5. The mathematical probability of 100,000 genes falling into place, and producing the specified complexity of trillions of human cells in correct sequence is nearly beyond calculation and reason. *(probabilities will be discussed more in chapter 6)*

6. As Richard Dawkins stated earlier, coded DNA sequencing is exactly analogous to our coded alphabet for language: [11, 20, 21]

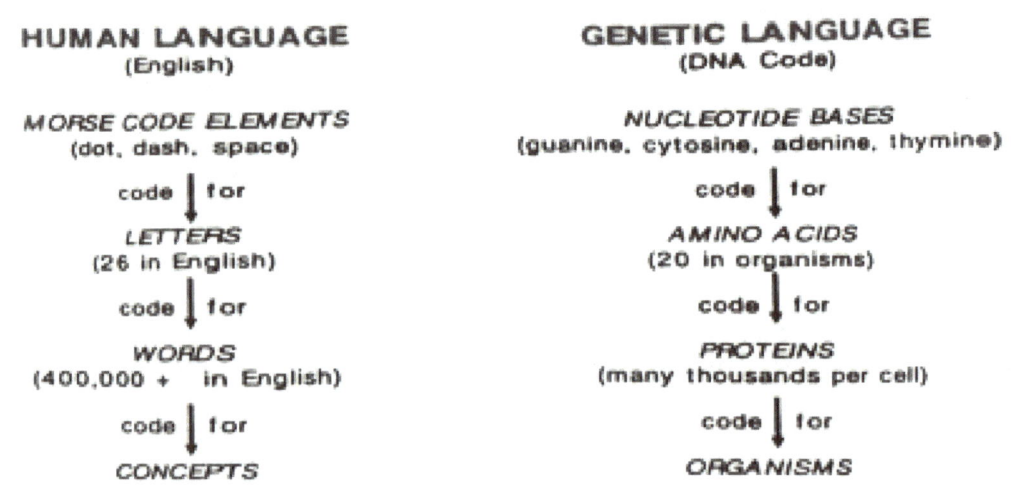

HUMAN LANGUAGE	GENETIC LANGUAGE
(English)	(DNA Code)
MORSE CODE ELEMENTS (dot, dash, space)	*NUCLEOTIDE BASES* (guanine, cytosine, adenine, thymine)
code ↓ for	code ↓ for
LETTERS (26 in English)	*AMINO ACIDS* (20 in organisms)
code ↓ for	code ↓ for
WORDS (400,000 + in English)	*PROTEINS* (many thousands per cell)
code ↓ for	code ↓ for
CONCEPTS	*ORGANISMS*

The analogy between human language and genetic language. The similarity is great enough to conclude that an intelligent cause was necessary for the origin of the DNA code. (Drawn by Kurt P. Wise.)

Veteran Oxford and Aberdeen professor, and life-long atheist Antony Flew, summarizes how genetic information played a huge role in reversing his scientific world view in his new book:

"There Is A God: How the World's Most Notorious Atheist Changed His Mind", (2007):
"What I think the DNA material has done is, that it has shown by the almost unbelievable complexity of the arrangements which are needed to produce it (life), that intelligence must have been involved in getting these extraordinarily diverse elements to work together. It's the enormous complexity of the number of elements and the enormous subtlety of the ways they work together. The meeting of these two parts at the right time by chance is simply minute. It is all a matter of the enormous complexity by which the results were achieved, which looked to me like the work of intelligence." [10]

Imagine Professor Flew's dilemma. He has spent his entire 50-year collegiate career espousing atheism at the world's most elite universities. He often dominated his opponents during debates as he promoted scientific atheism and Darwinism. Then at the apex of his career, Dr. Flew was faced with the realization that science was disproving his entire world view. Few scientists would have the courage to admit that everything their life had stood for was wrong, yet Dr. Antony Flew has done just that, and he is not alone. Many former naturalists are now accepting intelligence as the most logical creator of biological information, not random collisions of atoms. Dr. Stephen C. Meyer's new book, "Signature in The Cell: DNA and the Evidence for Intelligent Design", makes a powerful scientific argument for the presence of intelligence in genetic codes.

Flew, Antony (former atheist), and Roy Varghese., *"There is a God: How the World's Most Notorious Atheist Changed His Mind"*, HarperCollins Books, 2007.
"It now seems to me, that the findings of more than fifty years of DNA research have provided materials for a new and enormously powerful argument to design."

Gitt, Werner., Director at the German Federal Institute of Physics and Technology, *"In the Beginning was Information"*, Christliche Literatur-Verbreitung Publishing, 1997, p. 106.
"There is no known law of nature, no known process and no known sequence of events which can cause information to originate, by itself, in matter."

Tipler, Frank., professor of mathematical physics, *"The Physics of Christianity"*, New York: Doubleday, 1994.
"When I began my career as a cosmologist some 20 years ago, I was a convinced atheist. I never in my wildest dreams imagined that one day I would be writing a book purporting to show that the central claims of Judeo-Christian theology are in fact true, that these claims are straightforward deductions of the laws of physics as we now understand them. I have been forced into these conclusions by the inexorable logic of my own special branch of physics."

Yockey, Hubert P., Physicist; Army Pulse Radiation Facility, Aberdeen Proving Ground, Maryland, USA, *"Information Theory and Molecular Biology"*, Cambridge University Press: Cambridge UK, 1992, p. 7.
"It is perhaps clear to the reader that the genetic system is, in principle, isomorphic with communication systems designed by communications engineers. As a matter of fact, genetical systems have historical priority since organisms have been using the principles of information theory and coding theory for at least 3.8 x 10⁹ years!"

Lipson, H.S., "A Physicist's View of Darwin's Theory", *Evolutionary Trends in Plants*, **vol. 2, no. 1, 1988, p. 6. (2 quotes)**

"Design is the word that springs to mind on this subject. My biologist colleagues do not like it. They say that I should not object to a theory unless I have a better scientific one to replace it. … My unscientific theory is, that we have been designed in a macromutational way by an external creator. All the evidence supports this view but, of course, it cannot be sustained scientifically."

Denton, Michael., *"Evolution: A Theory in Crisis"*, **London: Burnett Books, 1985, p. 334.**

"The capacity of DNA to store information vastly exceeds that of any other known system; it is so efficient that all of the information needed to specify an organism as complex as man weighs less than a few thousand millionths of a gram. The information necessary to specify the design of all the species of organisms which have ever existed on the planet, a number according to G.G. Simpson of approximately one thousand million, could be held in a teaspoon, and there would still be room left for all the information in every book ever written!"

Cohen, L., Researcher and Mathematician, Member NY Academy of Sciences, Officer of the Archaeological Inst. of America, *"Darwin Was Wrong- A Study in Probability"*, **New Research Publications, 1984, p. 4.**

"At that moment, when the DNA/RNA system became understood, the debate between Evolutionists and Creationists should have come to a screeching halt!"

THE IMPORTANCE OF *'IRREDUCIBLE COMPLEXITY'*

Professor Flew is just one drop in a tidal wave of naturalist defections. These defectors are recognizing that genetic programming must first code for complex organic proteins, and then use them to construct highly complex organic 'machines'. Most of these organic machines will only function if all parts are simultaneously present, and this concept has been called *'irreducible complexity'*. [2, 3] Irreducible complexity is when a life-sustaining biological mechanism is composed of multiple, complex, interacting parts, all of which are necessary for that device to function. Removal of any one part would cause that organic machine to fail, and thus the organism to die. Since gradual evolution of each piece separately would not permit life, this forces one to recognize that each part of the complex organic machine would be forced to genetically mutate *simultaneously*. That makes intelligent intervention not just a possibility, but a virtual necessity. Nowhere on earth have mechanical machines self-assembled without an intelligent agent, and biology has numerous examples of irreducibly complex organic machines. One heavily researched, well-documented, and heavily debated example is the flagellum motor, seen on the next page. The flagellum motor is the 'propeller' that provides life-giving locomotion to thousands of species of microorganisms. All of the components of the flagella are necessary for it to function, which eliminates slow and gradual evolution as a potential creative agent. This genetic argument has been understood for decades, and only increases in strength as additional organic machines are discovered.

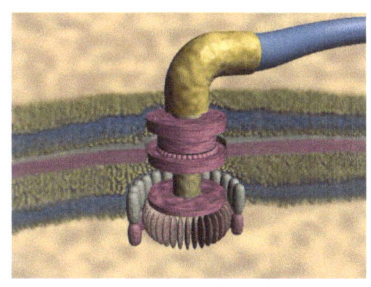 **Flagellum motor**

 **Mt. Rushmore**

(These graphics from www.ideacenter.org)

Namba, Keiichi., Professor of Frontier Biosciences at Osaka University, "Revealing the Mystery of the Bacterial Flagellum", *Nanonet,* **Interview, March 25, 2003.**

"The flagellum motor is a macromolecular assembly made of approximately 20 different proteins. It spans across three layers of membranes, namely, the cytoplasmic membrane, the peptidoglycan layer, and the outer membrane from the bottom to the top. It consists of various components such as a rotor, stators, a drive shaft, a bushing, a rotation-switch regulator, and so on."

Stumph, Michael., *Science News,* **vol. 173, June 7, 2008, p.18.**

"It's much, much more than just the organization of protein interactions. There's so much we don't know."

Peterson, Dan., 'The Little Engine that Could… Undo Darwinism", *The American Spectator,* **August 2005, p. 2.**

"All of the essential parts must be there, all at once, for the (flagellum) motor to perform its function of propelling the bacterium through liquid. Why is that important? Because that is precisely what Darwinian evolution cannot accomplish. Darwinian evolution is by definition "blind." It cannot plan ahead and create parts that might be useful to assemble a biological machine in the future. For the machine to be assembled, all or nearly all of the parts must already be there and be performing a function. Why must they already be performing a function? Because if a part does not confer a real and present advantage for the organism's survival or reproduction, Darwinian natural selection will not preserve the gene responsible for that part. In fact, according to the Darwinian theory, that gene will actually be selected against."

Sarfati, Jonathan., "Design in Living Organisms", *TJ,* **12(1):3–5, April 1998.**

"The famous British evolutionist (and communist) J.B.S. Haldane claimed in 1949 that evolution could never produce 'various mechanisms, such as the wheel and magnet, which would be useless 'till fairly perfect.' Therefore such machines in organisms would, in his opinion, prove evolution false. These molecular motors have indeed fulfilled one of Haldane's criteria. Also, turtles and monarch butterflies which use magnetic sensors for navigation fulfill Haldane's other criterion. I wonder whether Haldane would have had a change of heart if he had been alive to see these discoveries. Many evolutionists rule out intelligent design 'a priori', so the evidence, overwhelming as it is, would probably have no effect."

Carroll, Robert., Evolutionary Scientist/Naturalist, *"Patterns and Processes of Vertebrate Evolution",* **Cambridge University Press, 1997, p. 9.**

"How can we explain the gradual evolution of entirely new structures, like the wings of bats, birds, and butterflies, when the function of a partially evolved wing is almost impossible to conceive?"

Salisbury, Frank B., "Doubts About the Modern Synthetic Theory of Evolution", *American Biology Teacher,* **vol. 33, September 1971, p. 336.**

"Now we know that the cell itself is far more complex than we had imagined. It includes thousands of functioning enzymes, each one of them a complex machine itself. Furthermore, each enzyme comes into being in response to a gene, a strand of DNA. The information content of the gene, its complexity, must be as great as that of the enzyme that it controls."

CHAPTER 5:   IS THE EARTH REALLY 4.6 BILLION YEARS OLD?

The topic of earth age is critical to the Darwinian Theory, and vicious academic battles have been fought between 'old-earth' and 'young-earth' proponents.   Darwinism demands extremely long time periods, because untold trillions of 'helpful' genetic mutations would be necessary to macroevolve a single cell into the millions of life forms that have ever existed. Science has measured the rate that positive and negative mutations currently occur *(to be discussed in detail in the mutation section)* and it is apparent that even *billions* of years is not nearly enough time to generate complex life forms. Therefore, a young earth would be a certain deathblow to Darwinism. There are dozens of measuring techniques available for estimating the earth's age, but Darwinists only recognize the methods that suggest billions of years in age. They immediately ridicule and discard any dating techniques that suggest younger ages. Remember, science can only study systems that are observable, testable, and repeatable. Every dating technique used to calculate earth age breaks all three parameters, especially if the earth dates back millions or billions of years. The older the earth is, the more uncertain scientists are about what processes actually formed and shaped it. There are two opposing schools of thought when it comes to dating the earth, and each measures geologic time very differently.

1. *'UNIFORMITARIANS'* assume that ***"the present is the key to the past"***, that the earth's layers of crust accumulated throughout history at the same slow speed that they are accumulating today. Most geology books agree that the current earth sedimentation rate is approximately 15 cm per 1000 years. The 4.6 billion-year earth age, or "clock", is based on this theoretical aging logic and has set the time standard for many decades. This clock is 100% accurate as long as the earth has never suffered any catastrophic sedimentation events. As historically stated in geology textbooks:

Dunbar, Carl., Naturalist/Darwinist, *"Historical Geology"*, 2nd ed., New York, 1960, p. 18.
   "This philosophy (uniformitarianism) demands an immensity of time; it has now gained universal acceptance among intelligent and informed people."

2. Next there are *'CATASTROPHISTS'*. Catastrophists recognize that globe changing environmental events have been discovered throughout the earth's geologic history, like: super-volcanic eruptions, continental and/or global flooding events, ice ages, and massive meteor strikes. These planet altering cataclysms would have drastically affected the rate of sediment formation, as well as radiometric decay speeds. Catastrophists were ridiculed and persecuted for decades by traditional 'old-earth' scientists because they dared to suggest that the earth might be only a fraction of its commonly stated age of 4.6 billion years. However, many catastrophic environmental events are now well documented in the fossil record, and several have been proven to have altered the geography of the globe causing multiple extinction events. Darwinists grudgingly concede this truth when pressed, but seldom address this issue voluntarily because they would be forced to admit that traditional earth ages could be completely in doubt. Listen to the geologic experts:

Chadwick, Arthur V., Associate Professor of Biology at Loma Linda University's Geoscience Research Institute, "Megabreccias: Evidence for Catastrophism", as seen on *http://www.grisda.org/origins/05039.htm*, January 2009.
   "The presence of various kinds of megabreccias in the geologic column, showing in some cases the transport of extremely large clasts, indicates energy levels on a scale that staggers our imagination. Their common occurrence

in major portions of the geologic column of some localities indicates significant catastrophic activity in the past not readily explainable in terms of contemporary processes."

Handwerk, Brian., "National Geographic Considers Flood Myths Worth Investigating", *www.news. national-geographic.com,* **May, 2002.**

"New finds support worldwide flood myths," says a news report on the National Geographic website. The article talks about widely-separated instances of sunken cities, one off the coast of Cuba, and one in India, apparently tied to local legends of large floods sent by angry gods. Scientists, historians, and archaeologists view many of these enduring tales as myth, legend, or allegoric tales meant to illustrate moral principles. Recent findings indicate that at least a few of them could be based on real floods that caused destruction on an enormous scale. The Cuban structures are underwater as deep as 2500 feet. The Indian site covers several square miles off the coast. Explorer Graham Handcock's initial reaction was disbelief: "I have argued for many years that the world's flood myths deserve to be taken seriously, a view that most Western academics reject. But here in Mahabalipuram we have proved the myths right and the academics wrong."

USGS ,"Utah Geological Survey Notes", **Grand Canyon National Park, 34(3), 1-3, September 2002.**

"Dams of volcanic rock laid across the Grand Canyon have burst repeatedly and catastrophically over the past million years -- most recently about 165,000 years ago -- carrying enormous onrushing floods and carving out much of the great landmark (the Grand Canyon) in the blink of a geologic eye, new research by U.S. Geological Survey and University of Utah geologists suggests. The findings tend to support other new data indicating the canyon's Inner Gorge may be no more than 700,000 years old, much younger than earlier estimates of 3 million to 5 million years, said Robert Webb, a research geologist with USGS."

Krause, Lisa., "Ballard Finds Traces of Ancient Habitation Beneath Black Sea", *www.news. nationalgeographic.com,* **September 13, 2000.**

"Off the coast of northern Turkey, 311 feet (95 meters) below the Black Sea, explorer Robert Ballard (famous oceanographer who discovered the Titanic, and warships Bismarck, Lexington, and PT-109…) has discovered remains of an ancient structure that was apparently flooded in a deluge of Biblical proportions. The find may lend credence to a theory that a Black Sea flood gave rise to the Noah story and other flood legends. Last year, Ballard and his colleagues found proof that a catastrophic flood inundated the Black Sea in the region north of Turkey. The place and date of the flood, which may have occurred around 5,500 B.C., correspond to the time and location of the Old Testament account of Noah. Radiocarbon dating and paleontological evidence from a sample of shells and sediment collected from the site suggested that a massive flood occurred about 7,500 years ago. However, carbon dating using marine life is notoriously vague."

Parry, Donald W., Assistant professor of Hebrew at Brigham Young University and member of the international team of Dead Sea Scroll translators, *Ensign,* **January 1998.**

"Yet uniformitarianism, a premise on which much of geologic science is based, is an idea, not a fact. Uniformitarianism cannot explain all of the oddities and anomalies about the earth."

Bak, Per., "How Nature Works", **New York: Springer, Verlag, 1996, pp. 18-19.**

"Since catastrophism smacks of creationism, it has been largely rejected by the scientific community, despite the fact that catastrophes actually take place!"

Ager, Derek V., "*The Nature of the Stratigraphical Record*", New York: John Wiley and Sons, 1993, pp. 68-69. (2 quotes)

"Uniformitariansm triumphed because it provided a general theory that was at once logical and seemingly 'scientific'. Catastrophism became a joke and no geologist would dare postulate anything that might be termed a 'catastrophe' for fear of being laughed at or, in recent years, linked with a lunatic fringe... But I would like to suggest that, in the first half of the last century, the 'catastrophists' were better geologists than the 'uniformitarians'."

Allmon, Warren., "Post-Gradualism", *Science*, vol. 262, October 1993, p. 122.

"...the last 30 years have witnessed an increasing acceptance of rapid, rare, episodic, and 'catastrophic' events (in geologic history)."

Sunderland, Luther D., "*Mass Extinction and Catastrophism Replace Darwinism and Uniformitarianism*", Contrast: The Creation Evolution Controversy, vol. 4, no. 2, 1986, pp. 1-2, p. 343.

"The scientific establishment's acceptance of worldwide catastrophism and mass extinction does not signify their abandonment of materialistic evolution. Neither has their grudging acquiescence to the fact that great catastrophes caused the deposition of many of the fossils, which forced them to consider that virtually no fossils are in the process of forming on the bottom of any lake or sea today. This is a verboten subject. When I asked the editors of several of the most prestigious scientific journals the reasons for this silence, I was met with more silence."

Conclusion: Massive environmental catastrophes have happened regularly throughout earth history, and these catastrophes have affected the accuracy of both the geologic column clock and the radiometric clock. How inaccurate did they make these two clocks? Geologists do not actually know, because they have no clear picture of the earth's catastrophic record previous to man's recorded history. The fossil record gives clues about some specific events, but time has effectively obscured any accurate picture of ancient geologic processes. The earth could be billions, or millions, or thousands of years old, depending on the measuring tool that is chosen. Let's briefly examine 21 geologic time-measuring techniques:

21 METHODS OF DATING THE EARTH AND FOSSILS:

1.   THE GEOLOGIC COLUMN is one of the two *"clocks"* accepted by Darwinists. This uniformitarian dating system assumes that the earth's sediment layers are billions of years in the making, and that their formation has been steady and unchanged. They use present day sedimentation rates, as well as the thickness of the earth's various sediment layers to calibrate the ages of each geologic layer. The organic fossils trapped in each layer, called *'index fossils'*, are assumed to be the same age as its strata. This dating logic is scientifically sound only if the earth's sediment buildup has been steady and unchanged for 5 billion years. Any sudden sedimentation events would destroy this dating 'clock' and make the earth appear much older than it really is. Many geologic clues exist that suggest massive sedimentation events.

THE GEOLOGIC COLUMN

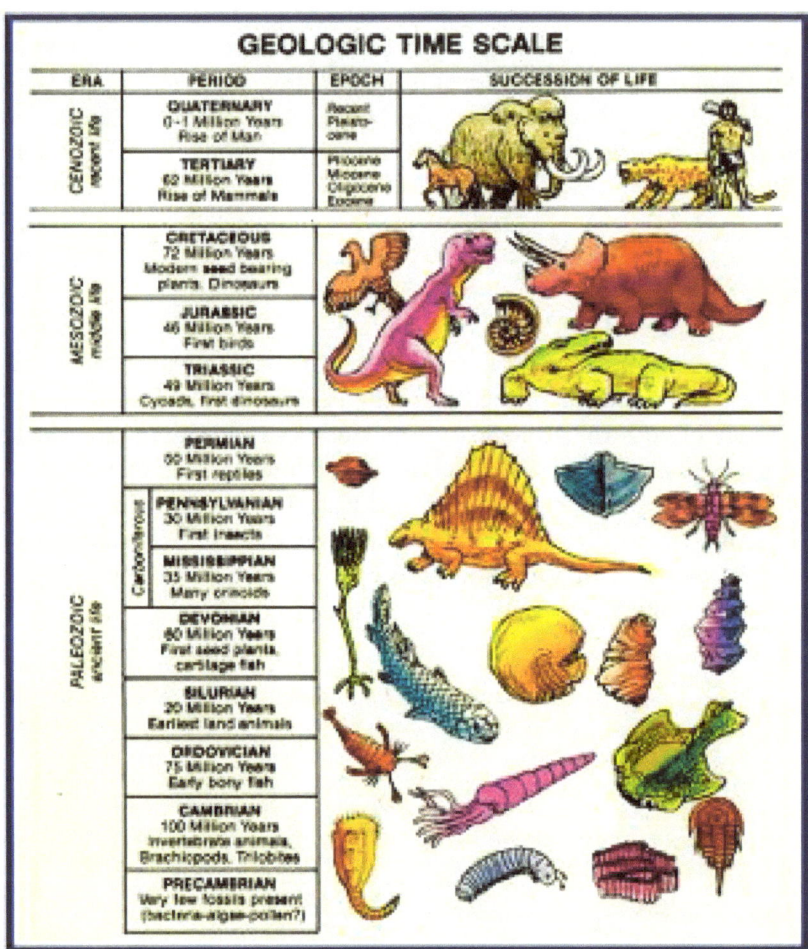

(Chart from www.nwcreation.net)

KNOWN PROBLEMS OF GEOLOGIC COLUMN DATING:

1. The 'geologic column' does not exist world-wide; it is a composite of many local columns.

2. Today's estimated sediment formation rate is roughly 15cm per 1000 years, so one way that evolutionists date the age of rock is by its depth and location between other sediment layers. Any sudden sedimentation events would destroy the accuracy of this dating logic.

3. The Standard *(imagined)* Geologic Column consists of 12 rock layers. Each is identified and dated by the index fossils trapped within, but the fossils are dated by the rock layer they are in. This is a classic example of unscientific circular reasoning.

4. *Assumed* Darwinian evolution ages have been assigned to the index fossils.

5. Radiometric dating formulas were programmed with 12 geologic assumptions. *(See next page)*

6. Fossils often appear in sediments as if pooled by catastrophic flooding. The Karoo Formation in South Africa, for example, is a mass grave that contains the fossils of over 800 BILLION vertebrates, including reptiles. *(Woodmorappe, John.,"The Karoo Vertebrate Non-problem: 800 Billion or Not", CEN Technical Journal, 14(2), 2000)*

7. There have been many strata-changing, catastrophic events recorded in geologic history.

8. There is no single, uniform, measurable, or predictable pattern to the world's strata layers.

9. Many strata are tightly bent into hairpin shapes without cracking, which suggests that the strata were formed quickly rather than slowly.

10. There are frequent "out-of-sequence fossils" appearing in contradictory strata.

11. The total absence of transitional fossils that bridge the gap between kingdoms and phyla is striking, and has led to the "punctuated equilibrium" theory. *(discussed in chapter ten)*

12. Dozens of equally scientific dating techniques contradict "uniformitarian" Earth ages.

13. Thousands of 'polystrate fossils' cross many geologic layers, scattered through 760 meters of strata, penetrating 20 geologic horizons. These fossils imply that the geologic layers were formed quickly, rather than slowly: [17, 27, 28, 29]

(Polystrate Tree Trunk, Internet Photo by: Ian Juby)

Huse, Scott M., "*The Collapse of Evolution*", 3rd Ed., Grand Rapids, MI: Baker, 1997, p. 96.
"Polystratic trees are fossil trees that extend through several layers of strata, often twenty feet or more in length. There is no doubt that this type of fossil was formed relatively quickly; otherwise it would have decomposed while waiting for strata to slowly accumulate around it."

2. RADIOMETRIC DATING is the second earth dating 'clock' accepted by Darwinists: All atoms **'decay'**, which means they continually lose energy and change structure as they age. For example, uranium decays into lead as the uranium atoms lose energy. The rate of decay is referred to as the element's **'half-life'**. The current decay speed for elements is known, but most elements are too stable to serve as research clocks, which means that they decay too slowly to be useful for 'atomic' dating. Only a few elements have a short enough half-life to be a prime candidate for measurable radiometric dating. To calculate a fossil or rock's age, one must examine the ratio of the original element in the fossil *(the parent element)*, to the element that it decays into over time *(the daughter element)*. The following four element pairs are the most commonly used radiometric dating clocks:

1. $C^{14} > C^{12}$ *('carbon dating')*

2. *Potassium> Argon*

3. *Uranium> Lead*

4. *Rubidium> Strontium*

If one knew the element's exact speed of decay, as well as the ratio of parent to daughter element, it is fairly simple to calculate the substance's age. The current decay speed for elements is generally agreed upon, but external environmental forces are known to alter those decay speeds. Over a dozen assumptions are unavoidable when creating decay speed formulas. If ALL of the below assumptions are guessed correctly, then radiometric dating would be a highly accurate method for determining the age of a sample being tested. But if ANY of the below assumptions are incorrect, then the calculated radiometric dates could be off by millions or billions of years:

ASSUMPTIONS INHERENT IN RADIOMETRIC DATING FORMULAS:

1. that only the "parent" element and none of the "daughter" element was originally present.
2. that the C^{14} ratio in the environment *(which would vary with land, plant, and animal ratios)* has remained perfectly constant throughout the earth's entire history, which scientists agree **has not been true.**
3. that none of the daughter element has moved into or out of the sample *(via gas escape)*.
4. that there is a sufficient quantity of the sample for an accurate test.
5. that solar explosions, solar flares, and magnetic reversals, which **are** happening and **do** affect the decay rates, have not affected the sample being tested.
6. that nuclear testing contamination has not affected the sample being tested.
7. that voltage input contamination has not affected the sample being tested.
8. that the earth's cosmic ray protection shield *(the natural magnetic field and ozone layer)* has never changed throughout the earth's entire history, **but evidence has proven that it has changed.**
9. that environmental chemicals and radiation eroding through the sample have not affected its decay process.
10. that it is contaminant free *(of carbon containing microorganisms, carbon from past fires…)*
11. that the traditional geologic column 'clock' is a proven fact, because these assumed geologic dates are used to verify and confirm index fossil dates.
12. that traditional Darwinian evolution dates are a proven fact. These hypothetical Darwinian time lines are used by labs to verify and confirm radiometric dates. [12, 17, 27, 28, 29]

HOW ROCK SAMPLES ARE NOT DATED:

Rocks are not dated by their appearance, they are not dated by their mineral content *(oil, coal, shale...)*, they are not dated by their structural features, they are not dated by adjacent rocks, and finally, they are not dated by any physical characteristics at all.

HOW ARE THEY DATED?

Rocks of unknown age are usually dated by their index fossils, and the age of the index fossils is derived from assumed evolutionary ages. This is incredibly unscientific circular reasoning.

DO ROCKS OF *KNOWN* AGE ALWAYS YIELD ACCURATE RADIOMETRIC AGES?

No. Many rock formations have "birthdates" that are known, like Sunset Crater in northern Arizona, Mt. Rangitoto in New Zealand, Vulcan's Throne in the Grand Canyon, the Kaupelehu Flow of the Hualalai Volcano, the Salt Lake Crater on Oahu, and Mt. Kilauea in Hawaii . When samples are tested radiometrically, they often give a range of dates that are *millions* or *billions* of years apart. When that happens, scientists simply choose a date that fits their presupposed supposition:

Geology Department Website, *http://www-geology.ucdavis.edu/,* **University of California-Davis, December 29, 2008.**

> *"… if you forget the idea of a strictly accurate molecular clock, and massage enough data, and rely on fossils for calibrating geological time, you can end up with something you believe. Molecular and fossil data can be 'partially reconciled', in the immortal phrase of the authors. (Douzery, et al. 2004 in: 'The Timing of Eukaryotic Evolution', Proceedings of the National Academy of Sciences)."*

Snelling, Andrew A. Ph.D., "The Fallacies of Radioactive Dating of Rocks Basalt Lava Flows in Grand Canyon", *www.wasdarwinright.com/dating.htm,* **September 5, 2006.**

> *"These radioactive dating methods have been used to calculate an absolute age of 1,103±66 million years for the Cardenas Basalt lavas. (The number after the ± symbol refers to the error margins in the "age" determination so that 1,103±66 million years means that the age is between 1,037 and 1,169 million years.) So it would seem! However, a closer examination of the results from all such studies reveals the fallacies of the radioactive dating methods. The claimed age of 1,103±66 million years was obtained using the rubidium-strontium isochron method with 10 samples and has been regarded as the best radioactive dating result for any Grand Canyon rock unit. Nevertheless, potassium-argon model "ages" for each of 15 individual Cardenas Basalt samples range from 577±12 to 1,013±37 million years, while the potassium-argon isochron "age" obtained using 14 samples is only 516±30 million years. This is less than half the rubidium-strontium isochron "age" of 1,111±81 million years obtained using 19 samples. It is also less than the claimed Cambrian age of the Tapeats Sandstone that sits on top of, and well above the Cardenas Basalt lavas. Worse still, the samarium-neodymium isochron "age" obtained using 8 samples is 1,588±170 million years—more than three times the potassium argon isochron "age" of 516±30 million years! So what is the correct "age" of the Cardenas Basalt lavas?"*

Surprisingly, even though radioisotope dating is often shown to be inaccurate when dating rocks of *known* age, I hear geologists constantly proclaim that it is accurate for dating rocks of an *unknown* age. When testing yields a range of dates for a rock or fossil, scientists often select an age that is most pleasing to their theory. The same inconsistencies are apparently true when carbon dating *organisms* of known and unknown age. Many researchers see the scientific hypocrisy here. Need a good example to verify this? Examine the radiometric date chart below, which was generated by NASA scientists after rock samples were returned and dated from the Apollo landings on the moon. Scientists had previously hypothesized that radiometric testing should yield lunar dates of 4.6 billion years, because that would coincide with contemporary theories of the day concerning the moon's age. Look at the huge inconsistencies of the laboratory dates. The ages given by radiometric testing ranged from .7 billion to 28.1 billion years old. How is it, that NASA confidently announced to the world that lunar rocks verified an age of 4.6 billion years old?! Because samples #15426 and 12013 yielded an age of 4.6 billion years old. The rest were ignored!

Apollo Sample No.	Ages in Billions of Years			Age inconsistencies, extremes in billions of years
	Uranium-Thorium-Lead Method		Potassium-Argon Method	
	Low	High		
10017	3.60	4.79	2.2	2.59
10057	3.96	4.17	2.3	1.87
10060	3.36	5.76	—	2.40
10084	4.31	8.20	>7	3.89
12070	3.63	4.50	>7	>3.37
12032	3.38	4.40	>7	>3.62
12063	3.75	4.09	2.6	1.49
12013	.7*	4.6	>6	>5.3
14310	5.3	11.2	—	5.9
14053	5.4	28.1	—	22.7
15426	4.6	16.2	—	11.6
66095	5.6	14.1	—	8.5

*Age determination using a Uranium-Thorium/Helium Technique

(NASA archives)

EXAMPLES OF "QUESTIONABLE" DATES, FOR VERY IMPORTANT FOSSILS:

1. In the 1970's, a key Darwinian primate skull, KNM-ER 1470, was initially tested and yielded an age of 212 to 230 M. *(million)* years, which, according to the Darwinian Theory, would be far too old to make geologic sense. In an attempt to get younger results, various other attempts were made to date the volcanic rock in the same "KBS-Tuff" area. After cataloging a list of rock dates, a date of 2.9 M. years was agreed upon. However, even this date was later determined to be "too old" to fit the Darwinian hypothesis. Eventually, a study of African pig fossils convinced anthropologists that skull 1470 must be much younger. A date of 1.9 M. years has now been officially accepted for KNM-ER 1470, which perfectly fits their original dating hypothesis. *(F.J. Fitch and J.A. Miller, "Radioisotopic Age Determinations of Lake Rudolf Artifact Site", Nature 226, April 18, 1970, p. 226 and Vincent J. Maglio, "Vertebrate Faunas and Chronology of Hominid-bearing Sediments East of Lake Rudolf, Kenya", Nature 239, October 13, 1972, pp. 379–85.) (Mehlert, A.W., "The Rise and Fall of Skull KNM-ER 1470, CMI, 1999.)* [17]

2. The dating of *Australopithecus ramidus*, another critical Darwinian fossil, was first determined to be 23 M. years old by the radiometric potassium-argon method. When it was decided that the test date seemed too old to fit the macroevolutionary scenario, 26 samples of surrounding basalt were tested, 17 of which gave a more "acceptable" age of 4.4 M. years. The other 9 samples yielded much older dates and were discarded. *(Lemonick, Michael D. and Andrea Dorfman, "One Giant Step for Mankind," Time, 158:54-61, July 23, 2001 and, G. Wolde Gabriel et al., "Ecological and Temporal Placement of Pliocene Hominids at Aramis, Ethiopia," Nature, 1994, 371:330-333.)*

3. All living organisms take in Carbon 14 (C^{14}) while they are alive, but stop absorbing it when they die. C^{14} is known to totally decay into Carbon 12 (C^{12}) in 50,000 years. Thus, any organic fossil that is thought to be older than 50,000 years should have no detectable C^{14} remaining. Coal, by traditional Darwinian geologic dating, should be at least 230 million years old and should, of course, have no remaining C^{14}. ***OF THE THOUSANDS OF SAMPLES OF COAL THAT HAVE BEEN CARBON DATED, EVERY SAMPLE TESTED HAS CONTAINED C^{14}.*** Darwinian paleontologists argue that the C^{14} in coal can't be original, because that would contradict the geologic column's clock. They contend that it must have entered as *contamination* from the earth layers surrounding the coal deposit. But wouldn't that contamination argument apply to ALL fossils and rocks, and thus destroy the accuracy of their ages as well? Contamination is a constant problem in "open" earth systems, and all fossils are buried in open systems. *(Hunt, Kathleen., "Carbon-14 in Coal Deposits", TalkOrigins Archive, May 22, 2002.)* [17]

4. Fossil wood found in "Upper Permian" rock, supposedly 250 M. years old, also still contained C^{14}. A sample of "Middle Triassic" wood, supposedly 230 M. years old contained C^{14}. All follow-up checks suggested no contamination, and the dates were deemed valid. *(A.A. Snelling, "Stumping Old-age Dogma". Creation, 1998, 20(4):48-50, and A.A. Snelling, "Dating Dilemma," Creation, 1999, 21(3):39-41.)*

The following experts in geologic and radiometric dating understand these problems very well, but they offer no tangible solutions. Geologic dating 'science' has remained un-changed for many decades, and the aforementioned problems have been recognized since its inception:

"KBS Tuff Shows the Flaws of Radiometric Dating", *Talk.Origins* website, 16 April, 2007.
"The lessons to be learned from the KBS Tuff dating controversy are not that radiometric dating does not work, but that it works with some caveats. It is those caveats that make radiometric dating totally dependent on uniformitarian geology and the geologic column."

Travis, John., "Microbes Muddle Shroud of Turin's Age," *Science News*, 147:346, 1995. *(2 quotes)*
"The reality of our biosphere is that virtually everything is permeated (contaminated) with microbes and their products. S.J. Mattingly and L.A. Garza-Valdes, of the University of Texas at San Antonio have been studying "biogenic varnishes" for years. These plastic-like coatings are produced by bacteria and fungi. …These biogenic varnishes may introduce carbon (C^{14}) that has been recently fixed from the atmosphere and thus make the sample's age appear younger than it really is."

Snelling, A.A., "U-Th-Pb Dating: An Example of False Isochrons", Proceedings of the Third International Conference on Creationism, Technical Symposium Sessions, 1994, pg. 503.
"Not only then has the open system behavior of these isotopes been demonstrated, but apparently 'isochrons' and

their derived 'ages' are invariably, geologically meaningless. Thus none of the assumptions used to interpret the U-Th-Pb radioactive system, used to yield 'ages', can be valid."

Eldridge, Niles., Naturalist/Darwinist, *"Time Frames: The Rethinking of Darwinian Evolution and the Theory of Punctuated Equilibria"*, NY: Simon and Schuster, 1985, pp. 51-52. (2 quotes)
"There is no way simply to look at a fossil and say how old it is unless you know the age of the rocks it comes from. …And this poses something of a problem: if we date the rocks by their fossils, how can we then turn around and talk about patterns of evolutionary change through time in the fossil record?"

Lee, R.E., "(Radiocarbon) Ages in Error", *Anthropological Journal of Canada*, vol. 19, No. 3, 1981, pp. 9, 29.
"The troubles of the radiocarbon dating method are undeniably deep and serious. Despite 35 years of technological refinement and better understanding, the underlying assumptions have been strongly challenged, and warnings are out that radiocarbon may soon find itself in a crisis situation. Continuing use of the method depends on a "fix-it-as-we-go" approach, allowing for contamination here, fractionation there, and calibration whenever possible. It should be no surprise, then, that fully half of the dates are rejected. The wonder is, surely, that the remaining half, come to be accepted. No matter how 'useful' it is, though, the radiocarbon method is still not capable of yielding accurate and reliable results. There are gross discrepancies, the chronology is uneven and relative, and the accepted dates are actually selected dates."

Azar, Larry., "Biologists, Help!", *Bioscience,* vol. 28, November 1978, p. 714.
"Are the authorities maintaining, on the one hand, that evolution is documented by geology and, on the other that geology is documented by evolution? Isn't this a circular argument?"

O'Rourke, J. E., "Pragmatism versus Materialism in Stratigraphy", *American Journal of Science,* vol. 276, January 1976, p. 53.
"The rocks do date the fossils, but the fossils date the rocks more accurately. Stratigraphy cannot avoid this kind of reasoning, if it insists on using only temporal concepts, because circularity is inherent in the derivation of time scales."

Welles, Samuel., two quotes from separate books: *"Fossils"*, World Book Encyclopedia, vol. 7, 1978 and *"Paleontology"*, World Book Encyclopedia, vol. 15, 1978. (*2 quotes*)
"Scientists determine when fossils were formed by finding out the age of the rocks in which they lie." (**And…**)
"The age of rocks may be determined by the fossils found in them."

O'Rourke, J.E., Evolutionist Researcher, "Pragmatism Versus Materialism in Stratigraphy", *American Journal of Science*, vol. 276, January 1976, pp. 47-55.
"The intelligent layman has long suspected circular reasoning in the use of rocks to date fossils and fossils to date rocks. The geologist has never bothered to think of a good reply, feeling the explanations are not worth the trouble as long as the work brings results. This is supposed to be hard-headed pragmatism."

Rastall, R.H., Lecturer in Economic Geology, Cambridge University, *Encyclopedia Britannica*, 1956, vol. 10, p. 1.
"It cannot be denied that from a strictly philosophical standpoint, geologists are here arguing in a circle. The

succession of organisms has been determined by the study of their remains imbedded in the rocks, and the relative ages of the rocks are determined by the remains of organisms they contain."

Conclusion: All Darwinian age measuring systems are dependent on each other for dating, all are based on unproven assumptions, none are known to be an independent and reliable dating technique, and all can give incorrect or multiple ages for newly created rocks or organisms. Dozens of other equally valid dating systems are rejected simply because their dates are *too young to fit the Darwinian Theory.* There are nearly two dozen additional methods for calculating the earth's age in addition to the geologic column and radiometric dating. Let's briefly examine these remaining dating techniques, each of which is based on measurable parameters. If we continue to use the Darwinian logic of: **"the present is the key to the past"**, these earth dating techniques should be every bit as valid as the geologic column and radiometric dating methods.

WARNING: **Darwinists severely ridicule the following dating methods and viciously attack anyone who dares to discuss them.** Why? Because they yield earth ages far younger than 4.6 billion years. These dating calculations suggest earth ages of only *millions* and/or *thousands* of years, either of which would be a deathblow to Darwinian macroevolution:

DATING METHODS THAT CONTRADICT DARWINIAN AGES

3. *SUN'S BURN RATE-* The current rate of the sun's fuel consumption is steady, measurable, and has a finite quantity *(assuming the present is the key to the past).* By measuring its decreasing diameter and solar neutrino data, the sun would have been so large only a few million years ago, that current Earth science theories would be in chaos. Calculating back BILLIONS of years would make the sun so large that it destroys all evolutionary concepts. [*New Scientist, vol. 97, March 1983*] [28, 29]

4. *LUNAR AGE DATA-* The moon's orbital effects heavily influence life sustaining factors on earth. The moon's origin and age is still completely unknown, but we know that it is steadily and measurably pulling away from the Earth as its orbit decays. Calculating back in time would find it *(assuming the present is the key to the past)* increasingly closer to Earth than it is today, but if it were too close, it would be pulled into the Earth by gravitational attractions. This physics limit is commonly known as the 'Rouche Limit'. When Rouche Limit calculations are combined with radiometric dates obtained from Apollo lunar rock samples *(wide range: .7 - 28.1 billion years old)*, and measurements of lunar surface dust, it suggests that the moon, and therefore all of Earth's biological systems, might be only millions of years old, not billions. [17, 30]

5. *FOSSIL WOOD* from Upper Permian rock layers, traditionally dated at 250 Million years old, was found to have Carbon-14 still present. Carbon-14 is known to disintegrate after 50,000 years *(assuming the present is the key to the past)*. These fossils cast extreme doubt on traditional geologic ages and suggest a much younger earth. *(A.A. Snelling, "Stumping Old-age Dogma". Creation, 1998, 20(4):48-50, and A. Snelling, "Dating Dilemma," Creation, 1999, 21(3):39-41.)*

6. *GALAXY/EARTH SPEEDS*- This is referred to as the "winding up dilemma". The stars of our own galaxy, the Milky Way, rotate about the galactic center with different speeds, the inner galaxies rotating faster than the outer and slowing down over time as energy dissipates. The current rotation speeds are known, and are so fast that *(assuming the present is the key to the past)* if our galaxy were more than a few hundred million years old, the increased speeds would cause a featureless smear of stars instead of its present spiral shape. The commonly accepted age of 4.6 billion years would result in nonsensically high rotation speeds. No such rotation dilemma exists if the galaxies are accepted as more recently formed. A similar argument can be made for the spin speed of the earth, *(currently rotating at 1,000 mph or 1,609 kph)* and slowing daily. Calculating back billion years would find the earth spinning so fast that centrifugal forces would cause it to flatten into a pancake. [*http://imagine.gsfc.nasa.gov/docs/ask_astro/answers/*]

7. *EARTH'S LAND EROSION*- Each year 25 billion tons of dirt and rock from the continents is deposited in the oceans. At today's rate *(assuming the present is the key to the past)* it would take only 15 million years to erode all land above sea level. Theories concerning 'land uplift' as it gets lighter from erosion, and calculations measuring this, are inadequate to compensate for all of this discrepancy. A similar argument can be made concerning the erosion of the earth's magnetic field, based on the rate of magnetic erosion seen today. [17]

8. *LAND AND SEA FLOOR SEDIMENT*- Classic geologic theories state that the ocean floors are 200 million years old, but based on the present rate of oceanic sedimentation *(assuming the present is the key to the past)*, the maximum age calculates to be less than 15 million years old. Not even the theories of subduction *(ocean floor pushing deeper into the earth)* could explain this discrepancy. Also there are large areas of sea floor *(Tasmanian Sea off Australian coast)* which cannot be part of such "subduction zones" and therefore make excellent measuring tools. Soft sediments from dead plants and animals form on the floor of the oceans at the rate of about one inch *(2.54 cm)* every 1,000 to 5,000 years. The depth of this 'ooze' suggests that the earth is quite young. Topsoil studies have shown that it takes 300 to 1,000 years to build one inch *(2.54 cm)* of topsoil *(assuming the present is the key to the past)*, yet the average depth of topsoil is about eight inches. [17, 28]

9. *OCEANIC SODIUM LEVELS*- Based on today's rate of sodium *(salt)* erosion *(assuming the present is the key to the past)* into the oceans *(450 million tons a year, 83% of which stays in the oceans)*, even if the oceans had no sodium to begin with, today's oceanic sodium level would accumulate in less than 42 million years... much less than the generally stated ocean age of 3 billion years. [17]

10. *FOSSILS THAT STRADDLE GEOLOGIC LAYERS*- There are many recorded instances where a single fossil, *(like a tree)*, straddles across several "million" years of geologic strata. These 'polystrates' could not happen under current 'old-earth' processes *(assuming the present is the key to the past)*, and the evidence points toward rapid burial and rapid sediment formation, rather than slow and gradual. Well-documented environmental catastrophes could easily explain these sudden sedimentation events, and sudden sedimentation would drastically reduce traditionally held theories on Earth age. [1, 12, 17]

11. *OUT-OF-SEQUENCE FOSSILS*- There are many recorded instances of "recently evolved" fossils being found in "ancient" strata *(assuming the present is the key to the past)*. One example is the well-documented pine tree pollen found in the Grand Canyon Precambrian Hakatai Shale. The complex Trilobite is a similar dilemma for Darwinists. Finds such as these cast added doubt on the geologic column's accuracy. [1, 12, 17]

12. *FOSSIL RADIOACTIVITY (RADIO-HALOES)-* These are rings of color formed around microscopic bits of radioactive minerals in rock crystals. They are observable and measurable fossil evidences of radioactive decay. Polonium-210 radio-haloes indicate that the Jurassic, Triassic, and Eocene formations in the Colorado Plateau were deposited within **months** of one another *(assuming the present is the key to the past)*, not hundreds of millions of years as the geologic timetable insists. [17, 28]

13. *ATMOSPHERIC HELIUM LEVELS-* All naturally occurring families of radioactive elements generate helium as they decay. By measuring the amount of helium in today's atmosphere *(assuming the present is the key to the past)*, and taking into account the known rate of helium escape into space, and assuming no helium was in the atmosphere to begin with, today's minute atmospheric helium levels would accumulate in less than 2 million years, which is only a fraction of the Darwinian atmospheric age. [17, 28, 30]

14. *QUANTITY OF HUMAN FOSSILS/ POPULATION STATISTICS-* Using modern, reliable, and scientific population formulas *(assuming the present is the key to the past)*, and when calculating for the supposed 1 million years of modern human activity on earth, there would have been 25,000 human generations, producing 10^{2100}, humans. *(Remember, 10^{80} is equal to the total number of atoms in the universe.)* When factoring in conservative death rate estimates and the fact that Neanderthal and Cro-Magnon man buried his dead along with artifacts, it can be shown with high reliability that the small number of human fossils found suggests that man has existed on earth for only thousands of years, not millions. [17]

15. *FOSSILIZATION PROCESS-* Organic tissues never fossilize *(assuming the present is the key to the past)* unless they are buried quickly and deeply in specific sediments before scavengers and bacteria can devour them. In addition, fossils are often found in *HUGE* groups suggesting burial by flooding. Their depths relative to each other often suggest hydrological sorting via sudden environmental catastrophe, not necessarily burial by body complexity over time. The very existence of the types and numbers of fossils in the geologic column also strongly indicates catastrophic conditions were involved in their burial and preservation. This evidence speaks strongly against slow and gradual strata layering, and speaks for sudden and massive catastrophic events... *which would destroy the reliability of the geologic column timetable.* [17]

16. *FOSSILIZATION TIME REQUIREMENTS-* It was always assumed that fossilization took millions/ billions of years due to the traditional Darwinian geologic column timetable. Experimentation has now proven otherwise. There is increasing proof via observation and experimentation that sedimentary rocks, coal deposits, oil pools, and bone fossilization are capable of forming in relatively short amounts of time *(assuming the present is the key to the past)* rather than the traditional Darwinistic timetables. For example, fresh lumbar has been successfully fossilized in 3 days under specific conditions. [28]

17. *DINOSAUR RED BLOOD CELLS, PRESERVED SOFT TISSUE,* and **HEMOGLOBIN** have been found in some unfossilized dinosaur bones. The maximum age for these tissues to remain intact *(assuming the present is the key to the past)* is conservatively estimated to be only a few thousand years, which strongly refutes the 65 million year age assigned to them by the geologic column. [*http://news.bbc.co.uk/2/hi/science/ nature/ 4379577.stm, March 2005*]

18. *THE DEPARTMENT OF ENERGY* drilled and examined radioactive zircon crystals extracted out of granite core samples taken from 15,000 feet deep in an undisturbed New Mexico granite formation. The presumed 1.5 billion year-old sample had lost essentially *none* of its radiogenic lead, even at the bottom of the hole where the temperatures were highest. Polonium radio-haloes were also found imbedded throughout the granite formation. Helium retention from the same samples *(197 degrees Celcius)* was amazingly high *(assuming the present is the key to the past)* and was, in itself, strong evidence for an earth age of less than 1 million years. The combined data would be consistent with an earth age in the *thousands* of years, as opposed to billion years. *[http://adsabs.harvard.edu/full/1980Metic..15..258.]* [17]

19. *GLOBAL EARTH CATASTROPHIES* continue to unfold in the Earth's geologic record. Great asteroids and super-volcanic eruptions are now proven, and are leading contenders as the catalysts for massive changes in Earth biology. Global flooding "myths" take on increased scientific credibility. Proven global catastrophes such as these should have *(assuming the present is the key to the past)* thrown traditional geologic time scales into chaos. [*Robert Ballard, www.NationalGeographic.com, fall of 2000*], [*www.news.national-geographic.com, May, 2002*]

20. *LACK OF 3RD STAGE SUPERNOVAS*- Astronomers indicate that supernovas go through three formation stages *(assuming the present is the key to the past)*. Based on the known size of the universe, and the classic "billions of years" age of the universe, there should be 35,000 third stage supernovas in our galaxy. It is conservatively predicted that 14% of them, or roughly 5,000, would be of a visible distance to earth. However, there are ***NO*** third stage supernovas in our galaxy, which is consistent with a universe age of less than 120,000 years, not the 15 billion years we are told today. [29]

21. *MORE CLUES TO EARTH AGE*- Over time, other evidence conflicting with the geologic timetable has been discovered: the ancient Mayan's witnessed living "ancient" archaeopteryx type birds, human artifacts and implements have been found embedded deep in undisturbed coal seams, pictographs of "long extinct" dinosaurs have been found drawn on cave and canyon walls, newly discovered living coelacanths, "ancient" wooly mammoths found instantly and freshly frozen, and "modern" fossils buried in "ancient" strata. Lastly, planetary angular momentum comparisons make little sense. According to naturalistic theories, the sun, which has 99.9% of the total mass of the solar system, should turn fastest, the planets somewhat slower, and their moons the slowest. The sun should have 700 times more angular momentum than all the planets combined, but in reality, the planets have 50 times more angular momentum than the sun. [*http://news.national geographic.com/news/2005/*] [17, 26, 28]

Conclusion: Examine the summarizing graph on the next page. It is my professional opinion, that *any* scientist, can present *any* of the preceding earth dating techniques, and make them to say *anything* they want them to say, to fit *any* hypothesis of their choosing. They are all supposedly based on measurable parameters but each method includes so many assumptions that scientists can make an argument for, or against, *any* and *all* of the earth dating techniques. So how old is the earth? Who knows. I personally place very little reliability in the traditional geologic times assigned to fossils. The only geologic comparisons that I feel are of any value are **relative** ages, not **literal** ages. In other words, fossils found near the surface of the earth are most certainly younger than those found deeper in the strata, but assigning specific ages to fossils and their corresponding strata seems slightly better than guesswork.

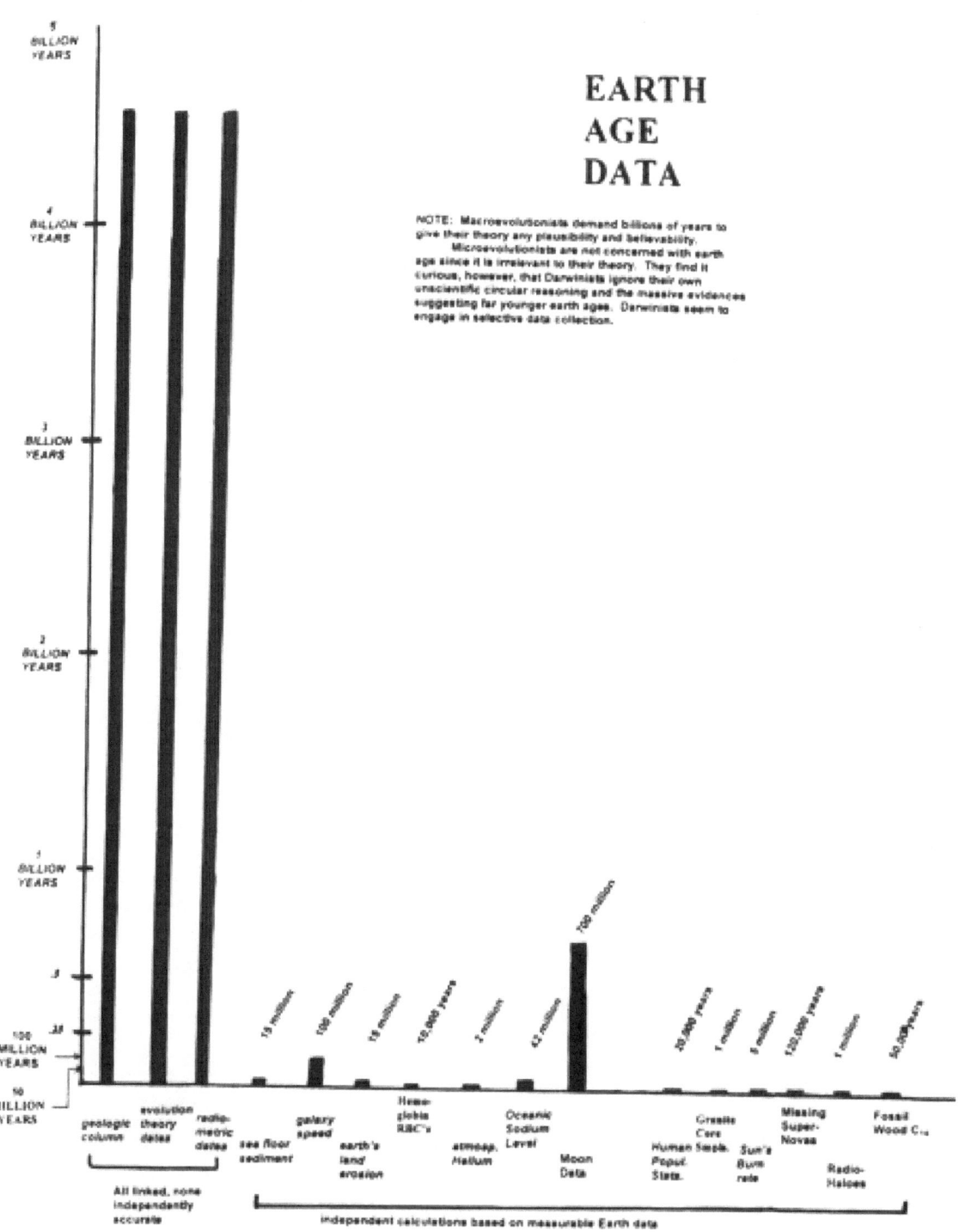

EARTH
AGE
DATA

NOTE: Macroevolutionists demand billions of years to
give their theory any plausibility and believability.
Microevolutionists are not concerned with earth
age since it is irrelevant to their theory. They find it
curious, however, that Darwinists ignore their own
unscientific circular reasoning and the massive evidences
suggesting far younger earth ages. Darwinists seem to
engage in selective data collection.

CHAPTER 6: WHAT MECHANISMS CAN CAUSE LIFE?

MECHANISM #1: SPONTANEOUS GENERATION

Earth life either assembled due to random chemical interactions *(accidental chance)*, or it was manufactured by intelligence *(planned and purposeful)*; those are the only two possible scenarios. Does any measurable evidence exist that suggests one is more likely than the other? Absolutely. Life is known to originate and reproduce only through *'biogenesis'*. This term means that life comes only from similar, pre-existing life. The *'modern cell theory'* fully confirms and compliments biogenesis, because it has shown that all life is made of cells, and cells are known to only originate from similar cells. All scientific evidence to date verifies these two fundamental biological truths. However, these facts completely contradict Darwinism, because Darwinism requires life to self-create from random, primordial, chemical interaction. *'Spontaneous generation'* is the old, disproved, scientific myth that once believed nonliving elemental matter could spontaneously organize into living organisms. This *'chemical evolution'* has been continuously and repeatedly disproven for over 150 years, and yet this is the very mechanism required by Darwinism to start life. To solve this dilemma, evolutionists created a new word for chemical evolution called *'abiogenesis'*. They insist that abiogenesis is the engine that drives the formation of organic life from raw elements, not spontaneous generation, but most bio-chemists agree that there is no functional or discernable difference between those terms. Abiogenesis is simply a re-label:

Internet Dictionary: *Merriam-Webster*, March 2009.
> *"ABIOGENESIS: the supposed spontaneous origination of living organisms directly from lifeless matter."*

Totten, R., "The Improbability of Abiogenesis; Spontaneous Generation Redux", *World-view Test Site*, 2003.
> *"A basic definition of 'abiogenesis' is: the chance origination of life from lifeless matter, totally through natural, unguided processes- which is essentially the same thing as 'spontaneous generation'."*

Science has clearly proven that life originating from inanimate chemicals is impossible, no matter what name you choose to call it. Life comes *only* from similar life forms, and cells originate only from like cells. And yet... ***Darwinism rejects the proven laws of science in favor of the disproved myth*** by insisting that all life forms on Earth came into being via spontaneous generation. This goes against all scientific observations and biological logic. Consider the following three scientific questions:

1. Could *100 trillion cells* self-organize into a human body, if given billions of years?
2. Could *1 cell* self-organize, by pure chance, from basic earth elements?
3. Could *1 functional protein molecule* self-organize, by pure chance?

Since the 1980's, scientists have been intensely studying the enormous conflict between *macroevolution* and *probability*. The gulf between what macroevolution demands from elements in order to create living cells, and what is mathematically possible given the incredible complexities of living systems, currently seems to be insurmountable. Probabilities are based on the total number of variables and their sequencing complexity. It is not difficult to calculate the probability of random chance forming various types of organic structures. *'Probability thresholds'* are used by scientists to determine the outer limit of what is scientifically possible due to chance. These thresholds can vary from researcher to researcher, but **1 X 10^{18}** is a conservative limit that encompasses most of the probability thresholds I have seen used over the years. Many scientists feel, even *that* number is way too generous, but I will use it for the sake of discussion. Compare that probability limit to the following scenarios, and it becomes clear why so many scientists have rejected Darwinism on this criteria alone:

MAJOR ELEMENTS IN A CELL (99.9%)	MAJOR MINERALS IN A CELL (.09%)	TRACE ELEMENTS IN A CELL (.01%)
==============	================	===================
Hydrogen	Calcium, magnesium	Iron, Selenium, Silicon
Oxygen	Phosphorus	Iodine, Molybdenum,
Carbon	Potassium, Sulfur	Copper, Fluorine, Manganese
Nitrogen	Chloride, Sodium	Zinc, Tin, Cobalt, Vanadium

1. COMPLEX ORGANIC MOLECULES: First, in order for spontaneous generation to occur, millions of the above **atoms** must be present, and self-organize by chance into the following highly complex molecular building blocks: CARBOHYDRATES, STARCH, GLUCOSE, AMINO ACIDS, PROTEINS, LIPIDS, NUCLEIC ACIDS, ETC. The simplest protein molecule contains 400 linked and sequenced amino acids. The probability of just 100 amino acids forming into a specific coded sequence by pure chance is **1 / 10^{53}**, and that is assuming that the amino acids first formed correctly by pure accidental chance. *(Remember, this already far exceeds the outer threshold of acceptance due to chance, 1 X 10^{18}).* This means that the odds of the simplest known protein, self-configuring, by chance, is far greater than what is scientifically acceptable. A conservative calculation for the smallest self-replicating **protein molecule** forming accidentally, by random chance, is **1/10^{450}**, which is why science has never witnessed this ever happening. [2, 3, 6]

Peet, J. H. John., BSc, MSc, PhD, CChem, FRSC, *www.truthinscience.org.uk,* **February 2009.**
> *"Consider a cell containing just 124 proteins. Professor Morowitz has calculated that the chance of all these forming without information input is 1 in 10^{100,000,000}. One of the smallest known genomes is that of Mycoplasma genitalium which manufactures about 600 proteins, so what are the chances of that happening without intelligent input? Humans have about 100,000 proteins."*

Aw, S., *CEN Tech. J.,* **Vol. 10, No. 3, 1996, p. 303.**
> *"The probability of forming the 2,000 or so enzymes needed by a cell lies in the realm of 1 in 1040,000. This makes the conceptual leap from even the most complex 'soup' to the simplest cell in the time available so dramatic that it requires some suspension of rationality in order to accept it."*

2. COMPLEX ORGANELLES: Those complex, non-living, organic molecules must defy all odds and *repeatedly* self-create, by the millions, and then further organize by chance into even higher functioning mechanisms called **organelles**: (CELL MEMBRANES, NUCLEAR MEMBRANES, CYTOPLASM, GOLGI, MITOCHONDRIA, RIBOSOMES, ENDOPLASMIC RETICULA, CHROMOSOMES, LYSOSOMES, CELL WALLS, CHLOROPLASTS, NUCLEI, ETC...) The calculated probabilities for spontaneously generating these organic machines would be thousands of times greater than the previous molecular calculation, which is why science has never witnessed this ever happening. [2, 3, 6]

3. *SELF-REPLICATING CELLS:* Those astonishingly complex organelles must finally, and immediately, self-organize into even higher complexity to form a living and self-replicating ***cell***. The probability of a living cell, with its millions of complex molecules and thousands of even more complex organelles, linking in sequence into an irreducibly complex life form, by pure chance, is conservatively estimated to be: **1 X 10 [4,478,276]**. Darwinists insist that spontaneous generation *(abiogenesis)* is the only plausible pathway to life. [2, 3, 6]

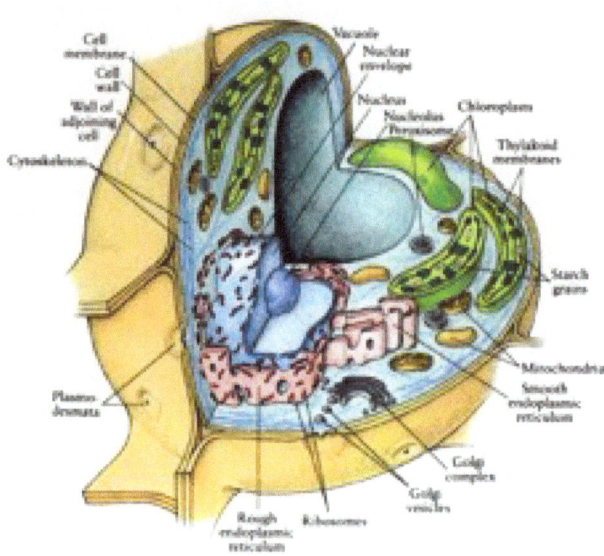

(Internet Photo by: www.daviddarling.info/images/)

Easterbrook, Gregg., "Where Did Life Come From?," *Wired Magazine*, February, 2007, p. 108.

> *"What creates life out of the inanimate compounds that make up living things? No one knows. How were the first organisms assembled? Nature hasn't given us the slightest hint. If anything, the mystery has deepened over time."*

Yockey, Hubert., *"Information Theory and Molecular Biology"*, Cambridge University Press, 1992, p. 257.

> *"The origin of life by chance in a primeval soup is impossible in probability, in the same way that a perpetual motion machine is in probability. The extremely small probabilities calculated in this chapter are not discouraging to true (evolutionary) believers. But a practical person must conclude that life didn't happen by chance."*

Hoyle, Sir Fred., physicist and professor of astronomy at Cambridge University, *Nature*, November 12, 1981, p. 148. (2 quotes)

> *"The likelihood of the formation of life from inanimate matter is one to a number with 40,000 zeros after it... It is big enough to bury Darwin and the whole theory of Evolution. There was no primeval soup, neither on this planet nor on any other, and if the beginnings of life were not random, they must therefore have been the product of purposeful intelligence."*

4. *COMPLEX ORGANISMS:* Lastly, consider the mathematical probability of a ***human body*** self-organizing over time, by random mutation *(birth defects)* and natural selection. This complexity nearly defies calculation. A human body is made of 100 TRILLION cells, and those complex cells are linked in sequence and order. Spontaneous generation has been thoroughly rejected by modern science as a life creating mechanism, and yet it is the very foundation that supports the entire Darwinian Theory. Scientists universally agree that it can not, and did not, create organic life:

McIntosh, A.C., Professor of Thermodynamics and Combustion Theory at the University of Leeds, Letter to: *The Daily Telegraph Newspaper*, March 2002.
"He (Professor Michael Behe) demolishes any possibility of Darwinian evolution at the biochemical level, and has masterfully shown there is irreducible complexity, down to minute details, in the workings of living organisms- details that will not go away with all the heat generated by professor Dawkins and his evident atheism."

Harold, Franklin M., *"The Way of the Cell: Molecules, Organisms and the Order of Life"*, Oxford University Press, 2001, p. 205.
"There are presently no detailed Darwinian accounts of the evolution of any biochemical or cellular system, only a variety of wishful speculations."

Bak, Dr. Per., "The Fifth Miracle", *New Scientist*, 160 (2155):47, October 1998.
"We are nowhere near understanding the origin of life."

Touchette, Nancy., "Evolution; Origin of Life", *Journal of NIH*, 5:95, 1993.
"So far, none of the current theories (of abiogenesis) have been substantiated or proven by experimentation, and no consensus exists about which, if any, of these theories is correct. Solving the mystery may indeed take longer than the origin of life itself."

White, Errol., Ichthyologist expert, *"A 1988 Proceeding of the Linnean Society"*, London: 177.8, 1988.
"I have often thought how little I should like to have to prove organic evolution in a court of law."

Denton, Michael Ph.D., Molecular Biologist, *"Evolution: A Theory in Crisis"*, Adler and Adler, 1985, pp. 264 and 324. (2 quotes)
"The complexity of the simplest known type of cell is so great that it is impossible to accept that such an object could have been thrown together suddenly by some kind of freakish, vastly improbable, event. Such an occurrence would be indistinguishable from a miracle."... "Even today we have no way of rigorously estimating the probability or degree of isolation of even one functional protein. It is surely a little premature to claim that random processes could have assembled mosquitoes and elephants when we still have to determine the actual probability of the discovery by chance of one single functional protein molecule."

Cohen, L., Researcher and Mathematician, Member NY Academy of Sciences, Officer of the Archaeological Inst. of America, *"Darwin Was Wrong- A Study in Probability"*, New Research Publications, 1984, p. 4.
"At that moment, when the DNA/RNA system became understood, the debate between Evolutionists and

Creationists should have come to a screeching halt!"

Wickramasinghe, Chandra., Professor of Astronomy and Applied Mathematics, Cardiff, and Sir Fred Hoyle Prof. of Astronomy, Cambridge, *"Evolution from Space"*, **J. M. Dent, 1981.** **(2 quotes)**
"Once we see, however, that the probability of life originating at random is so utterly minuscule as to make it absurd, it becomes sensible to think that the favorable properties of physics, on which life depends, are in every respect deliberate... It is almost inevitable that our own measure of intelligence must reflect higher intelligence, even to the limit of God."

Crick, Francis., co-discoverer of DNA, *"Life Itself: Its Origin and Nature"*, **New York: Simon and Schuster, 1981, p. 88.**
"An honest man, armed with all the knowledge available to us now, could only state that in some sense, the origin of life appears at the moment to be almost a miracle, so many are the conditions which would have had to have been satisfied to get it going."

Conclusion: Darwinian science collapses under the weight of biological spontaneous generation, and today's modern cell theory and biogenesis evidence all confirm that life only comes from similar, pre-existing life. Because life from abiogenesis has never been observed in the past or in the present, I am in total agreement with the scientists who insist that it is scientifically and mathematically impossible for a living cell to evolve randomly by pure chance. Today's high-tech microbiological sciences are incapable of assembling life from raw elements, and even if man does someday attain that ability it would only prove that *intelligent* forces could assemble a cell, not random forces. The mathematical complexities further intensify when one considers a myriad of other variables required by Darwinism. For example: male and female versions of millions of species of organisms must macroevolve with perfectly matched reproductive mutations *(birth defects)*, at exactly the same moment in earth history, and at exactly the same geographic location in order for reproduction to progress. This goes beyond logic and reason, and is completely void of scientific evidence.

But... haven't scientists like Oparin, Miller, Urey, and Haldane proven that life could have chemically evolved from primordial earth chemicals? **NO.** Their experiments were proven ***completely invalid*** decades ago. They used primordial chemicals in their tests that scientists now know were never present.

Moore, A., Naturalist/Darwinist, *Nature*, **453:31, 2008.**
"Speculations on the chemical origins of life are almost universally covered in school curricula under 'Evolution', despite the questionable relevance of the topic for evolution, and its rather uncertain scientific basis."

Orgel, L.E., Naturalist/Darwinist, *Scientific American*, **Vol. 271, No. 4, 1994, pp. 53-61.**
"Doubt has arisen (concerning abiogenesis) because recent investigations indicate the earth's atmosphere was never as reducing as Urey & Miller presumed."

Contemporary chemical evolutionists suggest that life may have formed on the 'backs of crystals', but they offer no measurable evidence to back this vague hypothesis. Others propose that the building blocks of life may have arrived by meteor, but recent studies show that organic matter would be heated and destroyed upon entering the earth's atmosphere. [9] Dr. Dean Kenyon, Professor Emeritus of Biology at San Francisco State University, was one of evolution's leading college text book authors. His books advocated the chemical evolution of life by abiogenesis. He has recently reversed his opinion and has concluded that chance interaction of chemicals could never create RNA, DNA, life proteins, or cells. This academic reversal was caused by the mounting evidence which confirms the reality that coded organic information is far too complex to assemble by chance. His scientific opinion changed because codes of this magnitude are only known to originate from intelligent sources. Darwinists once called Dr. Kenyon brilliant, but they now ridicule him. I have seen that to be the same fate for Antony Flew, and other Darwinists who have reversed their world view.

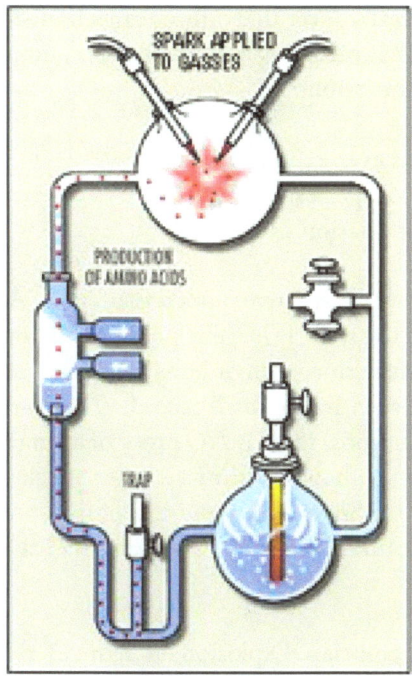

(Graphic from IdeaCenter.org)

The next problem for Darwinists to solve is, even if spontaneous generation did manage to do the impossible and actually created a living cell, how could that first primordial cell's genetic code increase in size and complexity in order to macroevolve into all of the life forms that have populated the earth throughout its entire history? In other words, how can a primitive 'bacteria' increase its genome to become a man, a mosquito, an elephant, and a tree? Darwinism suggests that mechanisms like variation, recombination, migration, isolation, and natural selection work together to slowly macroevolve organisms into new life forms. However, all of those mechanisms permit only microvariational change. None of them are capable of increasing the overall quantity of genes. Scientists have known for decades that only one mechanism has the ability to cause true genetic macroevolution, and that mechanism is *'mutation'*.

MECHANISM #2: MUTATION

"It must not be forgotten that mutation is the ultimate source of all genetic variation, that is, of material for natural selection."

> Mayr, Ernst., Naturalist/Darwinist, *"Populations, Species and Evolution"*, Cambridge: Harvard Press, 1970, p. 102.

Ernst Mayr's preceding quote is still absolutely true, and Darwinists universally agree that only one mechanism in nature is capable of creating increased quantities of DNA over time. That mechanism is *'mutation'*, more commonly known as, *'birth defects'*. So one must ask three important questions:

1. Can birth defects turn a primordial cell into every life-form that has ever existed?
2. Does the traditional geologic clock allow sufficient time to mutate all species?
3. Is there a trail of fossil evidence proving mutation actually caused speciation to occur?

It is common knowledge that mutations are random, rare, and nearly always destructive, not constructive. At best, mutations *reshuffle* existing genes, and usually *delete* genes, *reducing* the size of the gene pool. Mutation is therefore the worst mechanism for increasing the size, complexity, and function of an ordered genetic code. This process is totally inadequate for explaining the vast diversity of life seen in the fossil record. There are *NO* fossils of organisms proven to have intermediate and transitional mutations, there is *NO* proof of a single 'helpful' macromutation, or of any mutations greater than microvariational change within a known species. In fact, evolutionists are still struggling to define what a species actually is. So, even if hundreds of millions of years are assumed and allowed, mutations simply cannot create the massive levels of genetic increase needed to macroevolve all earth life from primordial cells.

Finally, when one closely examines the geologic chronology of the Cambrian Explosion, it reduces the time available for mutating the first primordial cells into *vertebrates (the most complex category of life)* to just a few million years. That is a mere blink of the eye on the Darwinian time scale. Random chance biological construction occurring at that speed would be somewhat analogous to building the Sears Tower from scratch in 2.5 *seconds*...while blindfolded! Darwinists recognize that this Cambrian time dilemma is another deathblow to their theory, so they are trying to find a mechanism that could cause *'hyper-evolution'* speeds. However, the probability problems caused by hyper-evolution are far worse than for slow, gradualistic evolution. Random mutation is the only naturalistic mechanism capable of increasing genetic codes over time, but it can not, and did not, macroevolve all earth species. Darwinian experts have been in agreement on this since the 1940's. Even prominent Darwinists like Stephan Jay Gould agree that mutation fails to create new species:

Gould, Stephen Jay., Leading Atheist/Darwinist and Professor of Geology and Paleontology at Harvard University, "Is a New and General Theory of Evolution Emerging", *Paleobiology*, vol. 6(1), January 1987.

> *"A mutation doesn't produce major new raw material (DNA). You don't make a new species by mutating the species."*

Hey, Jody., Department of Genetics at Rutgers University, "A Reduction of 'Species' Resolves the Species Problem", *http://lifesci.rutgers.edu/~heylab*, January 2009. (2 quotes)
> *"The species problem is the persistent biological and philosophical debate on the meaning of the word 'species' and the methods of species identification. …However, it also follows that many organisms do not belong to species. The finding that species exist, but that some organisms do not occur in species, reveals the central difficulty of systematic theories that assume the existence of species."*

Bataillon, Thomas., "Estimation of Spontaneous Genome-wide Mutation Rate Parameters: Whither Beneficial Mutations", *Heredity*, 25 December 2001, pp. 497-501.
> *"The actual rate of beneficial mutations is so extremely low as to thwart any actual measurement."*

Schwartz, Jeffrey H., Professor of Anthropology, University of Pittsburgh, USA], *"Sudden Origins: Fossils, Genes, and the Emergence of Species,"* John Wiley & Sons: New York NY, 1999, pp. 299-300.
> *"… Dobzhansky, as others did and would do, took for granted that, with enough time, the kinds of small mutations and changes that were observed in laboratory experiments on fruit-fly population genetics were also capable of producing the degrees of differences that seem to characterize species in the wild. To be sure, there was a certain logic in the belief that it was unnecessary to postulate another mechanism for evolutionary change when one already appeared to exist. This logic also seemed to benefit from the assertion that not only had no other mechanism (besides mutation) been observed but that no other mechanism had yet produced species. Nevertheless, it was and still is the case that, with the exception of Dobzhansky's claim about a new species of fruit fly, the formation of a new species, by any mechanism, has never been observed."*

Gerrish, PJ, and RE Lenski., "The Fate of Competing Beneficial Mutations in an Asexual Population", *Genetica*, 102/103, 1998, pp. 127-144.
> *"I have seen estimates of the incidence of the ratio of deleterious-to-beneficial mutations which range from one in one thousand, up to one in one million. The best estimates seem to be one in one million."*

Capra, Fritjof, *"The Web of Life"*, New York: Anchor Books, 1996, p. 228.
> *"It has been estimated that those chance (mutational) errors occur at a rate of about one per several hundred million cells in each generation. This frequency does not seem to be sufficient to explain the evolution of the great diversity of life forms, given the well-known fact that most mutations are harmful."*

Ohno, S., *"The Notion of the Cambrian Pananimalia Genome"*, Proc. Nat. Acad. Sci. USA, Vol. 93, August 1996, pp. 8475-8478.
> *"It follows that 6-10 million years in the evolutionary time scale is but a blink of an eye. The Cambrian explosion denoting the almost simultaneous emergence of nearly all the extant phyla of the Kingdom Animalia within the time span of 6-10 million years can't possibly be explained by mutational divergence of individual gene fluctuations."*

Koestler, Arthur., *"Janus: A Summing Up"*, Random House, New York, 1978, pp. 184-185.
> *"In the meantime, the educated public continues to believe that Darwin has provided all the relevant answers by the magic formula of random mutation plus natural selection, quite unaware of the fact that random mutations turned out to be irrelevant, and natural selection tautology."*

Grosse, Pierre-Paul., past-President, French Acadamie des Scientific, *"Evolution of Living Organisms"*, **Academic Press, New York, 1977, pp. 88, 97-98.**

> *"No matter how numerous they may be, mutations do not produce any kind of evolution. Mutations, in time, occur incoherently. They are not complementary to one another, nor are they cumulative in successive generations toward a given direction. They modify what preexists, but they do so in disorder...."*

Wills, Christopher., "Genetic Load", *Scientific American*, **vol. 222, March 1970, p. 107.**

> *"Any increase in the mutational load is harmful, if not immediately, then certainly to future generations."*

Martin, C., "A Non-Geneticist Looks at Evolution", *American Scientist*, **vol. 41, 1953, p. 101.**

> *"The truth is that there is no clear evidence of the existence of such helpful mutations. In natural populations endless millions of small and great genetic differences exist, but here is no evidence that they arose by mutation."*

Crow, James F., radiation and mutation specialist, "Genetic Effects of Radiation", *Bulletin of Atomic Scientists*, **Vol. 14, 1946, p. 1.**

> *"A random change in the highly integrated system of chemical processes which constitute life is certain to impair- just as a random interchange of connections in a television set is not likely to improve the picture."*

MECHANISM #3: VESTIGIAL ORGANS

Darwinists believe, and teach, that many species have *'vestigial organs'*, or, evolutionary left-over spare parts that are atrophying due to millions of years of lack of use or need. In other words, they are 'un-acquiring' certain characteristics. It is sort of an inverse *'Lamarckian acquired adaptation'* argument. Lamarck's disproven evolutionary theory originally suggested that physical traits developed during an organism's lifetime would be passed on to its offspring. For example: someone who takes up weight lifting will pass on larger muscles to his or her children, or if a person had a finger amputated, their children would be born without that same finger. The vestigial organ argument is the same logic, but in reverse. Research is now able to address two important points on the topic of vestigials: only the genetic code passes on physical characteristics, and useful functions are now known to exist for nearly all animal organs that were once believed unnecessary. During the past century the list of 180 human vestigial organs such as tonsils, appendix, adenoids, thymus gland, pituitary gland, and wisdom teeth dropped to virtually 0 by the year 2006. Both theories, Lamarckian adaptation and vestigial organs, are proven to be equally void of measurable scientific merit, and yet vestigials are still taught as truth in most biology textbooks and encyclopedias:

Spinney, Laura., Naturalist/Darwinist, "Vestigial Organs: Remnants of Evolution", *New Scientist*, **issue 2656, 14 May, 2008, p. 1.**

> *"It has even been suggested that the term (vestigial organ) is obsolete, useful only as a reflection of the anatomical knowledge of the day. In fact, these days many biologists are extremely wary of talking about vestigial organs at all."*

Tyler, David., Naturalist/Darwinist, as quoted in: "Vestigial Organs, Anyone?", *Uncommon Descent,* **26 November, 2007.**

> *"On this occasion (claiming the appendix was vestigial), the people with a design perspective were right and the Darwinians were wrong. Let's remember this next time we hear Creation or ID being decried as being unable to make any scientific predictions."*

Robinson, Tom., "Evolution's 'Vestigial Organ' Argument Debunked", *Good News Magazine,* **Fall, 2006, p. 1.**

> *"The tailbone, properly known as the coccyx, is another supposed example of a vestigial structure that has been found to have a valuable function. (Various) muscles attached to the tailbone are important for helping bowel and childbirth movements, for supporting internal organs and keeping the entrance of the alimentary canal closed. It also has an important function as a point of insertion for several muscles and ligaments, including the gluteus maximus, which is the large muscle that runs down the back of the thigh and allows us to walk upright."*

Perlof, James., *"Tornado in a Junkyard, the Relentless Myth of Darwinism"*, Refuge Books, May 1999.

> *"The thyroid gland, pituitary gland, thymus, pineal gland, and coccyx, once considered useless by evolutionists, are now known to have important functions. The list of 180 'vestigial' structures is practically down to zero. Unfortunately, earlier Darwinists assumed that if they were ignorant of an organ's function, then it had no function."*

Menton, David Ph.D., *"The Human Tail, and Other Tales of Evolution,"* St. Louis MetroVoice, Vol. 4, January 1994, No. 1.

> *"The appendix, like the once 'vestigial' tonsils and adenoids, is a lymphoid organ (part of the body's immune system) which makes antibodies against infections in the digestive system. Believing it to be a useless evolutionary 'left over,' many surgeons once removed even the healthy appendix whenever they were in the abdominal cavity. Today, removal of a healthy appendix under most circumstances would be considered medical malpractice."*

Enoch, H., *"Creation and Evolution"*, New York, 1966, pp. 18-19.

> *"Apes possess an appendix, whereas their less immediate relatives, the lower apes, do not; but it appears again among the still lower mammals such as the opossum. How can the evolutionists account for this?"*

Darwin, Charles., *"Origin of Species"*, 6th edition, London, 1872.

> *"There remains, however, this difficulty. After an organ has ceased being used, and has become in consequence much reduced, how can it be still further reduced in size until the merest vestige is left; and how can it be finally quite obliterated? It is scarcely possible that disuse can go on producing any further effect after the organ has once been rendered functionless. Some additional explanation is here requisite, which I cannot give."*

MECHANISM #4: EMBRYOLOGY

In contemporary high school and college biology classes, students are routinely taught that *'ontogeny recapitulates phylogeny'*, which means that the embryonic development of an individual organism *(its ontogeny)* follows the same path as the assumed evolutionary history of its species *(its phylogeny)*. Comparing the physical similarities of various types of unborn embryos *(such as fish, amphibians, reptiles, birds and mammals)* is known as *"comparative embryology"*. This concept was first pioneered by Ernst Haeckel of Germany well over 100 years ago. Haeckel was a contemporary of Charles Darwin. He examined, and drew pictures of a variety of embryos, and he theorized that the visual similarity of their embryonic formational stages was evidence for similar evolutionary ancestry. It was later discovered that he fraudulently falsified his pictures; his data is known to have no scientific basis whatsoever. He was convicted by an academic court in Jena, Germany and we now know that any slight visual similarity between embryos is merely superficial in nature. *(The red areas on the chart below are the key places that Haeckel faked)* Before Haeckel's conviction, Darwin contended that **embryology was the strongest evidence backing his theory!** So we now know that *"ontogeny does **NOT** recapitulate phylogeny"*. Embryonic recapitulation should be as rejected as Lamarck's disproven evolutionary theories, and while it is no longer used as evidence during serious academic evolutionary debates, it continues to be taught as truth in nearly all evolution text books.

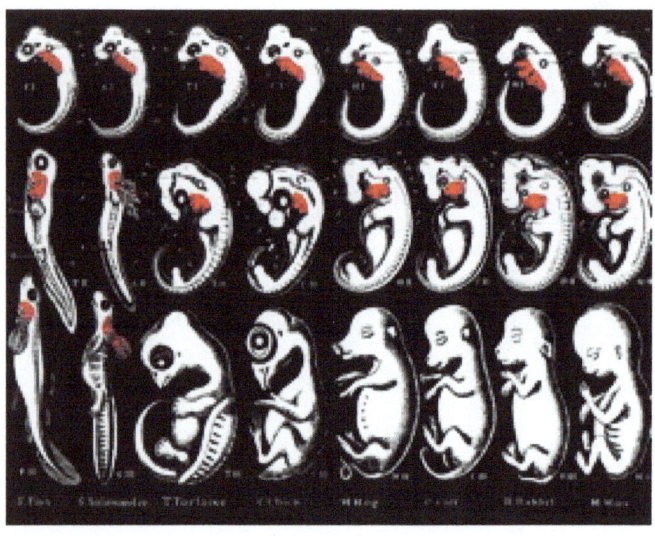

(This picture from :www.apologeticspress.org)

Maienschein, Jane., *"Evolution, Embryology and Ernst"*, **Arizona State University, Ludus Vitalis, Vol. XII, number 21, 2004, pp. 237-245.**

> *"Darwin considers a number of types of evidence weighing in favor of evolution, but he holds embryology as the strongest evidence of all."*

Richardson, Michael., "Haeckel's Embryos: Fraud Rediscovered", *Science*, **vol. 277, September 5, 1997, p. 1435. (2 quotes)**

> *"It looks like it's turning out to be one of the most famous fakes in biology… but Haeckel's confessions got lost after his drawings were subsequently used in a 1901 book called 'Darwin and after Darwin', and reproduced widely in English language biology text books."*

Milner, R., *"Encyclopedia of Evolution"*, Owlet Publishers, 1990, p. 206.
"When critics brought charges of extensive retouching and outrageous "fudging" in his famous embryo illustrations, Haeckel replied he was only trying to make them more accurate than the faulty specimens on which they were based."

Fix, William R., *"The Bone Peddlers: Selling Evolution"*, Macmillan Publishing Co., New York, 1984, p. 285.
"This idea (embryonic recapitulation) was fathered by Ernst Haeckel, a German biologist who was so convinced that he had solved the riddle of life's unfolding, that he doctored and faked his drawing of embryonic stages to prove his point."

Pitman, Michael., *"Adam and Evolution"*, Rider and Company, 1984, p. 120.
"To support his case, he (Haeckel) began to fake evidence. Charged with fraud by five professors and convicted by a university court at Jena, he agreed that a small percentage of his embryonic drawings were forgeries; he was merely filling in and reconstructing the missing links when the evidence was thin, and he claimed unblushingly that hundreds of the best observers and biologists lie under the same charge."

Gould, Stephen Jay., Naturalist/Darwinist, *"Ontogeny and Phylogeny"*, Harvard University Press, 1977, p. 430.
"He (German scientist Wilhelm His) accused Haeckel of shocking dishonesty in repeating the same picture several times to show the similarity among vertebrates as early embryonic stages in several plates of (Haeckel's book)."

Danson, R., *New Scientist* 49:35, 1971.
"Can there be any other area of science, for instance, in which a concept as intellectually barren as embryonic recapitulation could be used as evidence for a theory?"

MECHANISM #5: NATURAL SELECTION

This is commonly referred to as: *'survival of the fittest'*. Natural selection theology dictates that the strongest, smartest, and most adaptable organisms have the best chance of survival and reproduction. I completely agree that this makes perfect biological sense. All of the text books that I have read during my teaching career have taught that natural selection *'chooses'* and propagates off-spring with the most beneficial physiology based on the current environment, but there is one huge fact that evolution text books either overlook, or refuse to address. Natural selection alone *cannot increase the overall quantity of genes*, so it cannot lead to any form of macroevolution whatsoever. Natural selection is totally dependent on mutation to generate macroevolutionary change, and mutation has been proven incapable of creating any significant genetic improvements. Researchers have been in agreement on this for decades:

Kenneally, Christine., "Freedom from Selection Lets Genes Get Creative", *New Scientists*, issue 2675, 25 September 2008, p. 2.
"(Terry) Deacon may be right, that relaxed (natural) selection played an important role in human evolution,

but it's hard to imagine what evidence we could bring to bear on this (says anthropologist Richard Klein from Stanford University in California)."

Kelly, Kevin., Executive Editor of Wired Magazine, *"Out of Control: The New Biology of Machines",* **Fourth Estate: London, 1995, p. 475.**

"Despite a close watch, we have witnessed no new species emerge in the wild in recorded history. Also, most remarkably, we have seen no new animal species emerge in domestic breeding. That includes no new species of fruit flies in hundreds of millions of generations in fruitfully studies, where both soft and harsh pressures have been deliberately applied to the fly populations to induce speciation. And in computer life, where the term "species" does not yet have meaning, we see no cascading emergence of entirely new kinds of variety beyond an initial burst. In the wild, in breeding, and in artificial life, we see the emergence of variation. But by the absence of greater change, we also clearly see that the limits of variation appear to be narrowly bounded, and often bounded within species."

Lewin, Roger., *Science,* **217:1239-1240. 1982. (2 quotes)**

"Natural selection, a central feature of neo-Darwinism… may have a stabilizing effect, but it does not promote speciation. It is not a creative force as many people have suggested."

Patterson, Dr. Colin., Senior Paleontologist, British Museum of Natural History, London, *Interview,* **BBC television, March 4, 1982.**

"No one has ever produced a species by mechanisms of natural selection. No one has ever gotten near it…"

Koestler, Arthur., *"Janus: A Summing Up",* **Random House, New York, 1978, pp. 184-185.**

"In the meantime, the educated public continues to believe that Darwin has provided all the relevant answers by the magic formula of random mutation plus natural selection, quite unaware of the fact that random mutations turned out to be irrelevant, and natural selection tautology." *('tautology' means: useless repetition without any meaning.)*

Matthews, Harrison L., D.Sc., FRS, Naturalist/Darwinist, *"Intro to Origin of Species",* **Dent Publishers: London, 1972, p. 1.**

"The peppered moth experiments beautifully demonstrate natural selection or survival of the fittest. But they do not show evolution in progress. For however the population may alter in their content of light, intermediate or dark forms, all the moths remain from the beginning to end, (the species) Biston betularia."

MECHANISM #6: HOMOLOGY

Homology is the comparison of similar body structures. Darwinists insist that similarly constructed organisms are a key "proof" of gradual macroevolution over time. In fact, nearly all evolutionary fossil "chains" are nothing more than an extremely broad series of vaguely similar fossils. They are usually found chronologically in the geologic column, but never with any true transitional fossils to connect them. One classic *'biological descent with modification over time'* analogy used in science curricula to exemplify homology, is the analysis of automobile body style changes:

Berra, T., Evolutionist, *"Evolution and the Myth of Creationism"*, 1990, pp. 117-119.

> *"If you compare a 1953 and 1954 Corvette, side by side, then a 1954 and 1955 model, and so on, the descent with modification is overwhelmingly obvious. This is what paleontologists do with fossils, and the evidence is so solid and comprehensive that it cannot be denied by reasonable people."*

Darwinists contend that continuous, minor physiological change over short time frames will add up to major structural changes after eons of time *(continued microevolution = macroevolution)*. What the above Corvette analogy fails to mention is, Corvettes are complex machines designed and created by intelligence! The first Corvette did not just burst into existence spontaneously, and each model year's new variation required additional intelligent design… which also *"cannot be denied by reasonable people"*. This Corvette homology example perfectly demonstrates the naturalistic biases inherent in the organic homology argument.

Any group of fossils can be sorted according to increased complexity. One typical homological series presented in most textbooks is that of animal arms. The argument contends that if different animal arms have a similar bone structure, then that's proof of their evolutionary ancestry. The following diagram is a typical homological comparison shown in most evolution textbooks:

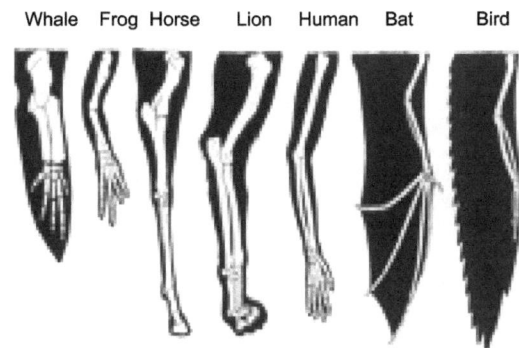

Whale Frog Horse Lion Human Bat Bird

(Internet Homology Diagram: www.cbu.edu)

However, any series of objects, organic or man-made, can be sorted homologically. Suppose we were to organize four buildings; a birdhouse, a doghouse, a tool shed, and a 3-bedroom house, from simplest to most complex? Would that serve as proof that the least complicated structure was created by spontaneous generation, and that each gradually 'evolved' into buildings of higher and higher complexity over time?

Of course not. Just like the Corvette example, it would only prove that the buildings have some physical similarities. Similarly, the homological grouping of organic fossils can not address these three important issues:

1. The organism's method of origin,
2. The potential presence of external design, and
3. The lack of transitional fossils connecting the series.

For decades the 'horse series' was presented in evolution text books as the best example of homologically related ancestors, but the evidence was so weak that it has now been eliminated from most evolution text books and museums. New DNA studies are also suggesting that homologous organs are usually produced by totally different gene complexes in each different species. This suggests that similar genes were not handed down from a common ancestor. If this is eventually confirmed, it would prove to be an additional deathblow to the homology argument. Increasing numbers of scientists are agreeing that homology is nothing more than a fossil grouping technique, and that it is an irrelevant proof of macroevolutionary change:

Meyer, Stephen C. Ph. D., Cambridge University, holds degrees in History, Philosophy, Physics and Geology, "Teach the Controversy", *Cincinnati Enquirer*, **March 30, 2002.**

> *"Shouldn't students know that many scientists doubt that the overall pattern of fossil evidence conforms to the Darwinian picture of the history of life? Shouldn't they know that some scientists now question previously stock Darwinian arguments from embryology and homology? And shouldn't they also know that many scientists now question the ability of natural selection to create fundamentally new structures, organisms, and body plans? Last fall 100 scientists, including professors from institutions such as M.I.T, Yale and Rice, published a statement questioning the creative power of natural selection. Shouldn't students know why?"*

Holden, Constance., "When is a Mandrill Not a Baboon?", *Science*, **vol. 283, 12 February 1999, p. 931.**

> *"(Their report) is a case study in how evolution can dupe casual observers- building (homological) similarities into unrelated species, and surprising differences into close cousins."*

Eldredge, Niles., as quoted in: Luther D. Sunderland, *"Darwin's Enigma: Fossils and Other Problems"*, **fourth edition, Master Book Publishers, Santee (California), 1988, p. 78.**

> *"I admit that an awful lot of that (imaginary stories) has gotten into the textbooks as though it were true. For instance, the most famous example still on exhibit downstairs (in the American Museum) is the exhibit on horse evolution prepared perhaps 50 years ago. That has been presented as literal truth in textbook after textbook. Now I think that that is lamentable...".*

Nelson, Gareth., Chairman and Curator of the Department of Herpetology and Ichthyology at the American Museum of Natural History, New York, interview with Tom Bethell, Wall Street Journal, December 9, 1986.

> *"We've got to have some ancestors. We'll pick... those. Why? Because we know they have to be there, and these are the best candidates (homologically). That's by and large the way it (Darwinism) has worked. I am not exaggerating."*

Fix, William., *"The Bone Peddlers: Selling Evolution"*, **Macmillan Publishing Company, New York, NY, 1984, p. 189.**

> *"Homologous organs are now known to be produced by totally different gene complexes in the different species. The concept of homology, in terms of similar genes handed on from a common ancestor, has broken down."*

Conclusion: None of the six mechanisms previously mentioned (*abiogenesis, mutation, vestigial organs, embryology, natural selection, and homology*) can produce life or macro-evolve life. Yet Darwinists insist that life spontaneously appeared from primordial earth chemicals and gradually mutated into all life forms. So even though there appears to be no mechanism to start life or to macro-evolve it, the proof of Darwinism will ultimately come down to... ***the fossil record***.

CHAPTER 7: THE ULTIMATE EVIDENCE... FOSSILS.

Examine the graph below. It organizes the known fossil record from the least complex life group *(bacteria)* to the most complex group *(chordates)* along the top from left to right, and then shows the corresponding geologic time scale vertically along the left side. According to the slow and gradual macroevolutionary theory, the simplest life group should first appear deepest in the fossil record if it is the earliest life form, with each life group of increased complexity appearing "higher" in the sediment as geologic time becomes more recent. But what does the fossil record actually indicate? It shows that 10 of the 15 main life groups originated almost *simultaneously* at the Cambrian layer *(which is the seldom taught **"Cambrian Explosion"**)*. The most complex life group, Chordates, first appears nearly as deep in the sediment as the earliest single cells. According to Darwinism, Chordates should be a very 'recently' evolved group. In fact, the seven most complex life groups make their début at nearly the same evolutionary time as the first 'primitive' life-forms. The fossil record speaks for itself: most life groups appear simultaneously, fully formed, and with no chain of connecting transitional fossils.

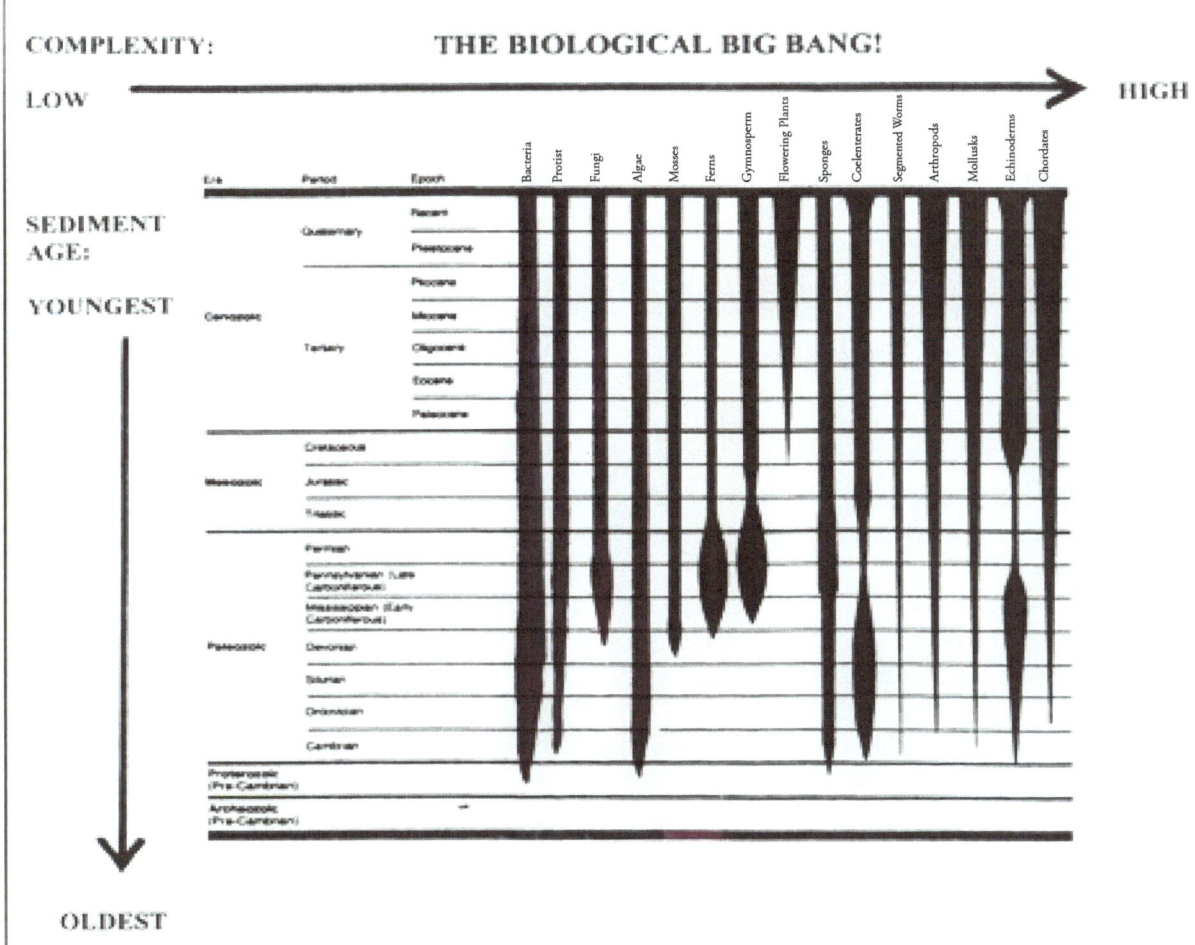

(Chart taken from reference # 5, p. 43)

HOW TO DEFINE A TRUE TRANSITIONAL FOSSIL

Darwinists and paleontologists both commonly agree that there are three requirements a fossil must meet in order to be considered a true macroevolutionary transitional fossil:

1. **Incipient *(transitional)* structures** such as half-scales/half-feathers on a "reptile/bird". *(Mosaics and homologies are never considered hard evidence for evolution, only incipients.)*
2. **A series of gradually changing intermediates** leading from one major classification group to another *(like from fish to amphibian)*, rather than sharp physiological changes.
3. **A chronological correlation** of those changes with the relative geologic time sequences.

(This is true...)

Stanley, Steven., Naturalist/Darwinist, *"Macroevolution: Pattern and Process"*, San Francisco: Freeman and Co., 1979, p. 2.

> *"...we must look to the fossil record for the ultimate documentation of large-scale change. In the absence of a fossil record, the credibility of evolutionists would be severely weakened. We might wonder whether the doctrine of evolution would qualify as anything more than an outrageous hypothesis."*

The above quote will always be scientifically accurate, because the credibility of any scientific theory is ultimately determined by measurable data, i.e. physical evidence. The fossil record is the physical evidence that will prove or disprove Darwinian macroevolution. Charles Darwin knew it, and today's naturalists know it. If gradual descent of all earth life by mutation and natural selection actually occurred, it will leave a specific type of fossil trail. What you are about to see is the actual fossil record. More and more paleontologists are realizing that the fossil record has become the most devastating evidence ***against*** Darwinian evolution:

(This is the reality of today's evidence...)

Barnes, R.S.K., and P. Calow, P.J.W. Olive, D.W. Golding, J.I. Spicer., *"The Invertebrates: A New Synthesis"*, 3rd Ed. (an invertebrate zoology text book), Blackwell Science Publications, 2001, pp. 9-10. *(2 quotes)*

> *"Most of the animal phyla that are represented in the fossil record first appear, fully formed and identifiable as to their phylum in the Cambrian some 550 million years ago... The fossil record is therefore of no help with respect to the origin and early diversification of the various animal phyla."*

Kemp, Dr. T.S., Naturalist/Darwinist and Curator of Zoological Collections, *"Fossils and Evolution"*, Oxford University, Oxford Press, 1999, p. 246.

> *"In virtually all cases, a new taxon appears for the first time in the fossil record with most definitive features already present, and practically no known stem-group forms."*

Knox, Barry, and Evans B., *"Biology"*, McGraw-Hill: Sydney, Australia, 1995, p. 663.

> *"Many people suppose that phylogeny can be discovered directly from the fossil record by studying a graded series of old to young fossils and by discovering ancestors, but this is not true. The fossil record supplies evidence of the geological ages of the forms of life, but not of their direct ancestor-descendant relationships There is no way of*

knowing whether a fossil is a direct ancestor of a more recent species, or represents a related line of descent (lineage) that simply became extinct."

Leyington, Jeffrey., "The Big Bang of Animal Evolution", *Scientific American,* **Vol. 267, November 1992, p. 84.**

"Evolutionary biology's deepest paradox concerns this strange discontinuity. Why haven't new animal body plans continued to crawl out of the evolutionary cauldron during the past hundreds of millions of years? Why are the ancient body plans so stable?"

Patterson, Dr. Colin., Senior Paleontologist, British Museum of Natural History, London, interview, BBC Television, March, 1982. *(2 quotes)*

"...Yet Gould and the American Museum people are hard to contradict when they say there are no transitional fossils... I will lay it on the line; there is not one such fossil for which one could make a watertight argument."

Hitching, Francis., *"The Neck of the Giraffe: Where Darwin Went Wrong",* **Penguin Books: New Haven, Ticknor and Fields, 1982, p. 19, pp. 56-57.**

"When you look for links between major groups of animals, they simply aren't there; at least, not in enough numbers to put their status beyond doubt. Either they don't exist at all, or they are so rare that an endless argument goes on about whether a particular fossil is, or isn't, or might be, transitional between this group and that. The curious thing is that there is a consistency about the fossil gaps; the fossils are missing in all of the important places."

Ridley, Mark., "Who Doubts Evolution?", *New Scientist,* **vol. 90, June 1981, p. 831.**

"In any case, no real evolutionist, whether gradualist or punctuationist, uses the fossil record as evidence in favor of the theory of evolution as opposed to special creation."

Woodruff, David S., "Evolution: The Paleobiological View", *Science,* **Vol. 208, 16 May 1980, p. 716.**

"Fossil species remain unchanged throughout most of their history and the record fails to contain a single example of a significant transition."

Kitts, David Ph.D., Naturalist/Darwinist, Head Zoology Curator, Dept. of Geology, *Evolution,* **vol. 28, September 1974, p. 467.**

"Despite the bright promise that paleontology provides a means of 'seeing' evolution, it has presented some nasty difficulties for evolutionists, the most notorious of which is the presence of 'gaps' in the fossil record. Evolution requires intermediate forms between species and paleontology does not provide them."

Researchers create scientific models to represent potential theories. They next gather data to see which model, if any, the evidence best supports. Currently, it appears that the fossil record provides very little evidential verification of macroevolution. In fact, the problem could be significantly worse for Darwinists. It appears that the fossil record is quite complete, and it suggests the OPPOSITE of what Charles Darwin predicted. Let's examine the two most commonly taught 'fossil-models' to see if today's fossil record supports either one.

MODEL A: **It is common knowledge among Darwinists, that if all life macroevolved via slow and gradual genetic mutation, one would expect to find:**

1. A clear geologic record of transitional fossils gradually evolving from simple to complex life,

2. A transitional fossil series chronologically linking all biological kingdoms, and...

3. No systematic gaps should exist anywhere in the fossil record. Gradualism should make biological classification, or taxonomy, virtually impossible to delineate as organisms slowly evolve. *(Darwin's "tree" of evolution)*

MODEL B: **On the other hand, if the major life groups suddenly and simultaneously appeared, and only microvariation has occurred within those groups, one would expect to find:**

1. Sudden fossil appearances in great variety, including complex forms *(Confirmed by the Cambrian Explosion)*,

2. No incipient transitional fossils would exist between the major taxonomic groups in the Earth's geologic record *(Confirmed by the fossil record)*,

3. Physiological characteristics would be fixed and complete *(Confirmed)*,

4. Wide classification gaps would exist, clearly separating taxonomic groups *(Confirmed)*, and...

5. Decay, destruction and slight microvariation over time should abound, not steadily increasing genetic complexity. *(Confirmed)*

Which model has been clearly visible in the fossil record for over **135 years?** .... Model B.   Darwin and modern geologic experts confirm this:

Darwin, Charles., *"Origin of Species"*, 6th edition, 1872, London, p. 413.
> *"The number of intermediate varieties, which have formerly existed on the earth, (must) be truly enormous. Why then is not every geological formation and every stratum full of such intermediate links? Geology assuredly does not reveal any such finely graduated organic chain: and this, perhaps, is the most obvious and gravest objection which can be urged against my theory."*

Koonin, Eugene., (National Center for Biotechnology Information), "Will Darwinism End With a Big Bang?", *Biology Direct*, October 8, 2007.
> *"Major transitions in biological evolution show the same pattern of sudden emergence of diverse forms at a new level of complexity. The relationships between major groups within an emergent new class of biological entities are hard to decipher and do not seem to fit the tree pattern that, following Darwin's original proposal, remains the dominant description of biological evolution.   The cases in point include the origin of complex RNA molecules and protein folds; major groups of viruses; archaea and bacteria, and the principal lineages within each of these prokaryotic domains; eukaryotic supergroups; and animal phyla In each of these pivotal nexuses in life's history, the principal "types" seem to appear rapidly and fully equipped with the signature features of the respective new level of biological organization.   No intermediate "grades" or intermediate forms between different types are detectable."*

Easterbrook, Gregg., "Where Did Life Come From?," *Wired Magazine*, February, 2007, p. 108.
"What creates life out of the inanimate compounds that make up living things? No one knows. How were the first organisms assembled? Nature hasn't given us the slightest hint. If anything, the mystery has deepened over time."

Mayr, Ernst., Naturalist/Darwinist, *"What is Evolution"*, Basic Books Pub., 2001, p. 189.
"Wherever we look at the living biota... discontinuities are overwhelmingly frequent. The discontinuities are even more striking in the fossil record. New species usually appear in the fossil record suddenly, not connected with their ancestors by a series of intermediates."

Pagel, Mark., "Happy Accidents?", *Nature,* Vol. 397, February 25, 1999, p. 665.
"Palaeobiologists flocked to these scientific visions of a world in a constant state of flux and admixture. But instead of finding the slow, smooth and progressive changes Lyell and Darwin had expected, they saw in the fossil record's rapid bursts of change, new species appearing seemingly out of nowhere and then remaining unchanged for millions of years... patterns hauntingly reminiscent of creation."

Carroll, Robert., Naturalist/Darwinist, *"Patterns and Processes of Vertebrate Evolution"*, Cambridge University Press, 1997, pp. 8-10.
"Fossils would be expected to show a continuous progression of slightly different forms linking all species and all major groups with one another in a nearly unbroken spectrum. In fact, most well-preserved fossils are as readily classified in a relatively small number of major groups..."

Eldredge, Niles., Naturalist/Darwinist, *"Reinventing Darwin: The Great Evolutionary Debate"*, Wiley Publishing, 1996, p. 95.
"No wonder paleontologists shied away from evolution for so long. It seems never to happen. Assiduous collecting up cliff faces yields zigzags, minor oscillations, and the very occasional slight accumulation of change over millions of years, at a rate too slow to really account for all the prodigious change that has occurred in evolutionary history. When we do see the introduction of evolutionary novelty, it usually shows up with a bang, and often with no firm evidence that the organisms did not evolve elsewhere! Evolution cannot forever be going on someplace else. Yet that's how the fossil record has struck many a forlorn paleontologist looking to learn something about evolution."

Futuyma, D. J., *"Science on Trial"*, New York: Pantheon Books, 1983, p. 197.
"Creation and evolution, between them, exhausts the possible explanations for the origin of living things. Organisms either appeared on the earth fully developed or they did not. If they did not, they must have developed from preexisting species by some process of modification. If they did appear in a fully developed state, they must indeed have been created by some omnipotent intelligence."

Gould, Stephen J., world renowned Darwinist, co-founder of the punctuated equilibrium theory, Professor of Geology and Paleontology at Harvard University, "Is a New and General Theory of Evolution Emerging?", *Paleobiology*, vol. 6 (1), 1980, pp. 119-130.
"The absence of fossil evidence for intermediary stages between major transitions in organic design, indeed our inability, even in our imagination, to construct functional intermediates in many cases, has been a persistent and nagging problem for gradualistic accounts of evolution."

Patterson, Dr. Colin., Senior Paleontologist of the British Museum of Natural History, *Interview*, April 10, 1979.

"If I knew of any (macroevolutionary transitions) fossils, nonliving or living, I would certainly have included them in my book. I will lay it on the line- there is not one such fossil for which one could make a watertight argument."

Raup, David., Curator of Geology at the Chicago Field Museum of Natural History, "Conflicts Between Darwin and Paleontology", *Field Museum of Natural History Bulletin*, Vol. 50, No. 1, Jan. 1979, p. 25.

"Darwin... was embarrassed by the fossil record... we are now about 120 years after Darwin and the knowledge of the fossil record has been greatly expanded. We now have a quarter of a million fossil species but the situation hasn't changed much."

Gould, Steven J., Darwinist/Atheist and Professor of Geology and Paleontology at Harvard University, "Evolution's Erratic Pace", *Natural History*, Vol. 86, No. 5, May 1977, pp. 12-16.

"The extreme rarity of transitional forms in the fossil record persists as the trade secret of paleontology. The evolutionary trees that adorn our textbooks have data only at the tips and nodes of their branches; the rest is inference, however reasonable, not the evidence of fossils.

Carter, G.S., Professor, Fellow of Corpus Christi College, Speech at Cambridge, England.

"We do not have any available fossil group which can categorically be claimed to be the ancestor of any other group. We do not have in the fossil record any specific point of divergence of one life form for another, and generally each of the major life groups has retained its fundamental structural and physiological characteristics throughout its life history..."

Conclusion: The problem for Darwinists is not that the fossil record has so few fossils that they can't yet prove their theory. The problem for Darwinists is far worse. As you just saw, the fossil record is quite complete and clearly shows the exact opposite of what Darwin predicted. It overwhelmingly proves that the foundational assumptions governing his entire theory were ***wrong***. And remember, Darwin himself recognized that the fossils required to prove his theory were missing. He simply thought that the transitional fossils would be found long after his death:

Darwin, Charles., *"Origin of Species"*, 6th edition, 1872, London, p. 413.

"The number of intermediate varieties, which have formerly existed on the earth, (must) be truly enormous Why then is not every geological formation and every stratum full of such intermediate links? Geology assuredly does not reveal any such finely graduated organic chain: and this, perhaps, is the most obvious and gravest objection which can be urged against my theory."

But the missing fossils have not been found. So today's biggest evidence question becomes: are there ***ANY*** macroevolutionary transitional fossils ***anywhere*** in the fossil record to prove that microorganisms became plants and invertebrates, which became vertebrates, which became amphibians, which became reptiles, which became birds, which became land mammals, which became marine mammals... and that primates became man? Let's examine each of these critical 'transitional' stages in detail.

If life began with 'simple' single-celled microorganisms as Darwinists assert, those cells eventually had to macroevolve into multicellular invertebrates. This would require a tremendous increase in genetic and physiological complexity, which would have created over-whelming fossil evidence. What is seen in the fossil record? There are no transitional fossils in the geologic column linking the earliest single-celled organisms to the complex invertebrates that supposedly arose from them. The oldest life-bearing fossil layer, the Cambrian Layer, shows a sudden and massive ***"Cambrian Explosion"*** of life *(the Biological Big Bang)*, where just before there was no life. Thousands of species of clams, snails, trilobites, sponges, brachiopods, worms, jellyfish, sea urchins, sea cucumbers, swimming crustaceans, sea lilies, and many other complex invertebrates make their first appearance, fully formed, at virtually the same moment as single celled organisms, and with no transitional fossils connecting them. A few Precambrian fossils have been discovered, but they too are complex invertebrates, not intermediate links from single-celled microorganisms. Darwinists commonly argue that the linking fossils do not exist because soft and delicate tissues easily decay and seldom fossilize. This argument holds no validity since bacteria and other "soft" microscopic creatures are found throughout the entire length of the geologic column. Transitions leading from primordial cells to invertebrates should be seen by the billions in the fossil record. [4, 12, 14]   The experts agree, there are none:

Meyer, Nelson, Chien, and Ross., *"The Cambrian Explosion: Biology's Big Bang"*, from: Darwinism, Design, and Public Education, (reference 6), 2003, p. 326.

> *"The suddenness of the appearance of animal life in the Cambrian, "the Cambrian explosion" has now earned titles such as "The Big Bang of Animal Evolution" (Scientific American), "Evolution's Big Bang" (Science), and the "Biological Big Bang" (Science News)."*

Kemp, Dr. T.S., Curator of Zoological Collections at Oxford University, *"Fossils and Evolution"*, Oxford UniPress, 1999, p. 246.

> *"In virtually all cases a new taxon appears for the first time in the fossil record with most definitive features already present, and practically no known stem-group forms."*

Douglas, Erwin, et al., "The Origin of Animal Body Plans", *American Scientist*, vol. 85, 1997, p. 126.

> *"All of the basic architectures of animals were established by the close of the Cambrian Explosion; subsequent evolutionary changes, even those that allowed animals to move out of the sea onto land, involved only modifications of those basic body plans."*

Eldredge, Niles., Naturalist/Darwinist, *"Reinventing Darwin: The Great Evolutionary Debate"*, Wiley Publishing, 1996, p. 95.

> *"No wonder paleontologists shied away from evolution for so long. It seems never to happen. Assiduous collecting up cliff faces yields zigzags, minor oscillations, and the very occasional slight accumulation of change over millions of years, at a rate too slow to really account for all the prodigious change that has occurred in evolutionary history. When we do see the introduction of evolutionary novelty, it usually shows up with a bang, and often with no firm evidence that the organisms did not evolve elsewhere! Evolution cannot forever be going on someplace else. Yet that's how the fossil record has struck many a forlorn paleontologist looking to learn something about evolution."*

Bengtson, Stefan., Naturalist/Darwinist, *Nature*, 345:765, 1990.

"If any event in life's history resembles man's creation myths, it is this sudden diversification of marine life when multicellular organisms took over as the dominant actors in ecology and evolution. Baffling to Darwin, this event (the Cambrian Explosion) still dazzles us and stands as a major biological revolution on par with the invention of self-replication and the origin of the eukariotic cell. The animal phyla emerged out of the Precambrian mists with most of the attributes of their modern descendants."

Sunderland, Luther., *"Darwin's Enigma: Fossils and Other Problems"*, 4th edition, Master Books, 1988, p. 9. (2 quotes)

"Now, after over 120 years of the most extensive and painstaking geological exploration of every continent and ocean bottom, the picture is infinitely more vivid and complete than it was in 1859. Formations have been discovered containing hundreds of billions of fossils and our museums now are filled with over 100 million fossils of 250,000 different species. The availability of this profusion of hard scientific data should permit objective investigators to determine if Darwin was on the right track. What is the picture which the fossils have given us? ...The gaps between major groups of organisms have been growing even wider and more undeniable. They can no longer be ignored or rationalized away with appeals to imperfection of the fossil record."

Dawkins, Richard., Darwinist/Naturalist, *"The Blind Watchmaker"*, New York, W.W. Norton, 1987, p. 229.

"It is as though they (the Cambrian fossils) were just planted there, without any evolutionary history. Needless to say, this appearance of sudden planting has delighted creationists."

Futuyma, D., *"Evolutionary Biology"*, Sunderland Mass.: Sinauer Assoc., Inc., 1986.

"It is considered likely that all the animal phyla became distinct before or during the Cambrian, for they all appear fully formed, without intermediates connecting one form to another."

Gould, Stephen J., Naturalist/Darwinist, "A Short Way to Big Ends", *Natural History*, v. 95, January 1986, p. 18.

"Where, then, are all of the Precambrian ancestors- or, if they didn't exist in recognizable form, how did modern complexity get off to such a fast start?"

Eldredge, Niles., and Ian Tattersall, Naturalist/Darwinist, *"The Myths of Human Evolution"*, Columbia University Press, 1982, pp. 45-46. (several quotes)

"Darwin himself, ... prophesied that future generations of paleontologists would fill in these gaps by diligent search ... One hundred and twenty years of paleontological research later, it has become abundantly clear that the fossil record will not confirm this part of Darwin's predictions. Nor is the problem a miserably poor record. The fossil record simply shows that this prediction is wrong. ...Paleontologists, faced with a recalcitrant record, obstinately refusing to yield Darwin's predicted pattern, simply looked the other way."

Eldredge, Niles., Paleontologist at the American Museum of Natural History, *"The Monkey Business: A Scientist Looks at Evolution"*, Washington Square Press, N.Y., 1982, p. 4. (several quotes)

"Beginning about six hundred million years ago... the earliest known representative of the major kinds of animals still populating today's seas made a rather abrupt appearance. This rather protracted "event" shows up graphically in the rock record all over the world, at roughly the same time, thick sequences of rocks,

barren of any easily detected fossils, are overlain by sediments containing a gorgeous array of shelled invertebrates: trilobites, brachiopods, mollusks.... Creationists have made much of this sudden development of rich and varied fossil record where, just before, there was none... Indeed, the sudden appearance of a varied, well preserved array of fossils... does pose a fascinating intellectual challenge."

INVERTEBRATES > VERTEBRATES

The evolutionary jump from invertebrates to backboned vertebrates involves thousands of complex physiological changes. The origin of vertebrate fishes, and afterward their even more complex land off-spring, would have been the greatest macroevolutionary event in Earth history. These gradually increasing physiological complexities should be clearly visible in the fossil record, and the strata should be full of transitional fossils documenting these changes in body forms. However, Darwinists are unable to identify which invertebrate group is the ancestor of vertebrates, because no transitional fossils have been found to connect any of the groups. One single fossil found in the Burgess Shale of Canada, *'Pikaia'*, is presented by Darwinists as the only intermediate to fill the entire 100 million year evolutionary gap, and even it is not considered to be a true incipient transitional fossil: [4, 12, 14]

"*Is Pikaia* an Arthropod?", Understanding Evolution, *http://evolution.berkeley.edu/evolibrary/article/* March 2009.

"Take a good look at the fossil and reconstruction of Pikaia (below). Then compare what you see to the arthropod checklist. Doesn't it look like Pikaia inherited all of the characteristics of an arthropod (invertebrate)?"

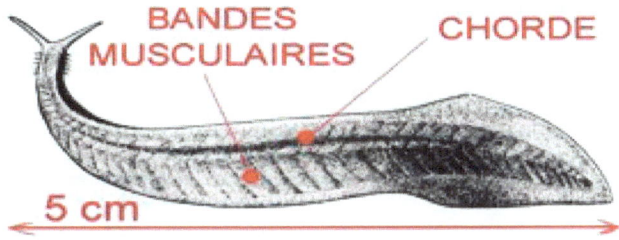

(Internet Pikaia Photo: http://pagesperso-orange.fr.)

***http://everything2.com/title/Pikaia%2520gracilens*, Dec. 05, 2002.**

"Found in the Burgess Shale, Pikaia gracilens was a worm-like fish measuring 3-4 centimeters that lived roughly 500 million years ago. Like all chordates, it had a dorsal nerve cord supported by a cartilage-like notochord that ran the length of its body. Though it is the earliest known chordate, there is no reason to believe that Pikaia is a direct ancestor of humanity. It was not a vertebrate, and is actually grouped as being an early member of the cephalochordata, a modern family of eel-like animals that includes lancelets and amphioxus. With fewer than 60 specimens identified and most of those in poor condition, opinion is still mixed on Pikaia's exact place in the tree of life."

There should be *millions* of intermediate transitional fossils connecting invertebrates to the first vertebrate fish groups since there are *billions* of well-preserved specimens on both sides of the invertebrate-vertebrate gap, but again, none exist:

Hickman, C.P., Roberts and Larson., Naturalist/Darwinist, *"Integrated Principals of Zoology"*, McGraw Hill, NY, 899, 2001, p. 511.
"...the first fishes left no fossil record and their form and relationships are a mystery."

Carroll, Robert L., Naturalist/Darwinist, *"Patterns and Processes of Vertebrate Evolution"*, Cambridge University Press, 1997, p. 296.
"We still have no evidence of the nature of the transition between cephalochordates and craniates. The earliest adequately known vertebrates already exhibit all the definitive features of craniates that we can expect to have preserved in fossils. No fossils are known that document the origin of jawed vertebrates."

Kuru, Mustafa., Naturalist/Darwinist, *"Omurgal Hayvanlar"* (Translation: "Vertebrates"), Gazi University Publications, 5th ed., Ankara, 1996, pp. 21, 27. *(2 quotes)*
"There is no doubt that Chordates evolved from invertebrates. However, the lack of transitional forms between invertebrates and Chordates causes people to put forward many assumptions. ...The views stated above about the origins of Chordates and evolution, are always met with suspicion since they are not based on any fossil records."

Bone, Q., with N.B. Marshall and J.H.S. Blaxter., *"Biology of Fishes"*, Blackie Academic & Professional, Glasgow:UK, 1995, p. 6.
"... we still know very little about the early origins of fishes."

Strahler, N., *"Science and Earth History: The Evolution/Creation Controversy"*, Prometheus Books, 1987, p. 405.
"In sediments of late Silurian and early Devonian age, numerous fishlike vertebrates of varied type are present, and it is obvious that a long evolutionary history had taken place before that time. But of that history we are mainly ignorant."

Todd, G. T., *American Zoology*, 20(4): 757, 1980, p. 757. *(2 quotes)*
"All three subdivisions of the bony fishes appear in the fossil record at approximately the same time. How did they originate? ...And why is there no trace of earlier intermediate forms?"

Norman, J. R., Dept. of Zoology, British Museum of Natural History, London, and Dr. P.H. Greenwood, British Museum of Natural History *"Classification and Pedigrees: Fossils, A History of Fishes"*, New York: Hill and Wang Publishing, 1975, p 343.
"The geological record has so far provided no evidence as to the origin of the fishes..."

Ommanney, F. D., *"The Fishes"*, **Life Nature Library, New York: Time-Life, 1964, p. 60.**

"How the earliest chordates evolved, what stages of development it went through to eventually give rise to truly fishlike creatures, we do not know. Between the Cambrian when it probably originated, and the Ordovician when the first fossils of animals with really fishlike characteristics appeared, there is a gap of perhaps 100 million years which we will probably never be able to fill."

--INSECTS--

Insects are the largest class in the animal kingdom, and like the invertebrates and vertebrates they explode onto the fossil scene suddenly and fully formed with *no* evolutionary link to another life group. The Pennsylvanian Period of the geologic column has appropriately been called: *"The Age of Insects"*. When one examines those earliest insect fossils, many are so well preserved that their sense organs are easily recognizable. Evolutionists hypothesized that the less complicated non-flying insects would have evolved into the more complex flying insects, but again, no intermediates have been found. "Wings" are irreducibly complex organic machines that just appear, fully formed and fully functioning in the fossil record, in all animal lines containing flight. The macroevolutionary example most commonly taught in textbooks is that of the peppered moth. While it provides an excellent example of micro-genetic variation within a species, the peppered moth is not evidence for macroevolution. [12, 14]

Tyler, David., "Stasis in the Fossil Record of Leaf Insects", *Science Literature Website*, **January 14, 2007. (2 quotes)**

"Here we report the first fossil leaf insect, Eophyllium messelensis gen. et sp. nov., from 47-million-year old deposits at Messel in Germany… This fossil leaf insect bears considerable resemblance to extant (existing) individuals in size and cryptic morphology, indicating minimal change in 47 million years. This absence of evolutionary change is an outstanding example of morphological and, probably, behavioral stasis (stasis: unchanging physiology)."

Encyclopedia Brittanica On-Line., *"Insect: Insect Fossil Record"*, **2001.**

"No (insect) fossils have yet been found from the Late Devonian or Early Carboniferous periods when the key characters of present-day insects are believed to have evolved; thus early evolution must be inferred from the morphology of extant (currently living) insects."

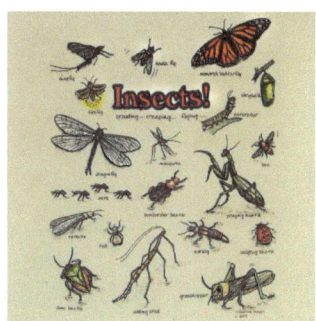

(Internet Photo from: http://images.google.com/images)

Wootton R. J., and C. P. Ellington, "Biomechanics and the Origin of Insect Flight", in *Biomechanics in Evolution*, Cambridge University Press, 1991, p. 99.

> *"When insect fossils first appear, in the Middle and Upper Carboniferous, they are diverse and for the most part fully winged. There are a few primitively wingless forms, but no convincing intermediates are known."*

Bernays, Elizabeth., University of Arizona, *"Evolution of Insect Morphology in Relation to Plants"*, 1991.

> *"The evolution of (insect) morphological adaptations is largely veiled in the past, and direct measures of any evolutionary change are rare."*

Hutchins, A.E., *"Insects"*, 1988, pp. 3-4.

> *"Insect origins beyond that point (the Carboniferous period) are shrouded in mystery. It might almost seem that the insects had suddenly appeared on the scene, but this is not in agreement with accepted (evolutionary) ideas of animal origins."*

Encyclopedia Britannica, Vol. 7, 1978 edition; *Macropaedia*, p. 585.

> *"The fossil record does not give any information on the origin of insects."*

FISH > AMPHIBIAN:

The body modifications leading from gill-breathing fish to lung-breathing land animal would require massive genetic alterations. There should be thousands of transitional fossil lines documenting this progression. Evolutionists have traditionally assumed that amphibians were the first **'tetrapods'** *(four-legged land animals)*, and also assumed that tetrapods evolved from a fish ancestor. Assumptions are necessary due to the total absence of transitional fossils. Ancient lobe-finned fish like the 'extinct' **Coelacanth**, once thought to be intermediate links from fish to amphibian, supposedly used their lobed fins to *"walk on the sea floor before crawling onto land."* Live Coelacanths, thought to be extinct for hundreds of millions of years, were discovered to be still living and thriving in 1938. They have proven to be modern, fully formed fish.

(Internet Coelacanth Photo from: http://www.g-3d.com/G-3d)

More than 4000 amphibian species have been identified. The "earliest" fossilized amphibian, *Ichthyostega*, is hardly transitional since it has fully formed legs, shoulders, and pelvic girdles. All fish commonly cited as being the most likely ancestors of amphibians are 100% fish, while all of the so-called descendant amphibians are 100% amphibian with their basic limbs, feet, and legs. [4, 12, 14] One relatively new fossil, *Panderichthys stolbovi*, is being proposed as a potential transitional fossil, but it has not yet been critically examined, and is far from having any incipient structures *(half feet-half fins, etc...)* so Darwinists, again, must rely on hypothetical fossils:

Biggs, Daniel, Lederman, Ortleb, Rillero, Zike at al., "*Life Science*", public school text book by: Glencoe McGraw-Hill, New York, NY, online at www.science.glencoe.com, 2009. *(2 quotes, underlines added)*
> *"Today's bony fish are probably descended from the first jawed fish...amphibians are thought to have evolved from these lobe-finned fish..."*

Daeschler, et al., Naturalist/Darwinist, "A Devonian Tetrapod-like Fish and the Evolution of the Tetrapod Body Plan", *Nature*, vol. 440:757-763, April 6, 2006.
> *"...the origin of major tetrapod features has remained obscure for lack of fossils that document the sequence of evolutionary changes."*

Carrol, Robert L., Naturalist/Darwinist, "*Patterns and Processes of Vertebrate Evolution*", Cambridge University Press, 1997, p. 230.
> *"...neither the fossil record nor study of development in modern genera yet provides a complete picture of how the paired limbs in tetrapods evolved."*

Helfman, G.S., with B.B. Collette and D.E. Facey. "*The Diversity of Fishes*", Blackwell Science, MA., 1997, p. 157.
> *"No intermediate fossils between jawed and jawless forms have been found - early fossils of jawed fishes had jaws, teeth, scales and spines. The origins of jaws and other structures that characterized the early gnathostomes are lost in the fossil record, belonging to some group about which we known nothing."*

Carroll, Robert L., Naturalist/Darwinist and Curator of Vertebrate Paleontology, Redpath Museum, McGill University, Montreal, Canada, "*Vertebrate Paleontology and Evolution*", W.H. Freeman & Co: New York NY, 1988, pp. 4 and 138. (2 quotes)
> *"Where information regarding transitional forms is most eagerly sought, it is least likely to be available. ...We have no intermediate fossils between rhipidistian fish and early amphibians."*

Forey, P. L., *Nature*, vol. 336, 1988, p. 727. (2 quotes)
> *"...there was then a long-held belief that coelacanths were close to the ancestry of tetrapods. ...But studies of the anatomy and physiology of Latimeria (the coelacanth) have found this theory of relationship to be wanting, and the living coelacanth's reputation as a missing link seems unjustified."*

Adler, Kraig., "*Encyclopedia of Reptiles & Amphibians*", Equinox, Oxford, 1986, p. 4.
> *"Although this transition (from fish to amphibian) doubtless occurred over a period of millions of years, there is no known fossil record of these stages."*

Taylor, Gordon Rattray., award winning science writer and former editor of the BBC's *"Horizon"* series, *"The Great Evolution Mystery"*, Harper and Row, 1983, p. 60.

"There are no intermediate forms between finned and limbed creatures in the fossil collections of the world."

AMPHIBIAN > REPTILE:

Darwinism teaches that amphibians macroevolved into reptiles. The fossils *Seymouria* and *Diadectes* are commonly taught to be the main amphibian to reptile transitions, but there is one huge problem… their ages are **backward.** Amphibians, which should be much older than reptiles, date back only 280 million years, while the earliest reptile fossils *Hylonomus* and *Paleothyris* date back much farther, 310 million years. This makes reptile 'descendants' 30 million years *older* than their supposed 'ancestors'. The widely divergent reptile groups include: the flying reptiles, the marine reptiles, the gliding reptiles, the snakes, and the turtles. No true transitional fossils have been discovered to bridge any of these reptile groups, nor are there any transitions to cross the huge gap from amphibians. The oldest known fossil from each reptile group is fully formed with no incipient structures. [4, 12, 14]   It is for these reasons that paleontologists and the fossil record speak loudly against amphibian to reptile macroevolution:

"Oldest Snake Fossil Shows a Bit of Leg", *New Scientist,* 19 April, 2006, p. 1.   *(3 quotes)*

"Zaher admits that … there are many questions about snake evolution left unsolved. At the top of that list is the question of what lizard group is most closely related to snakes. 'We do not have an undisputed hypothesis on that question' he admits. Michael Caldwell, from the University of Alberta, Canada argues '…without identifying a closest ancestor, there is no robust way of gaining insight into the origin of snakes'."

Carroll, Robert L., Naturalist/Darwinist, "The Origin and Early Radiation of Terrestrial Vertebrates", *Journal of Paleontology*, no. 6, November 2001, pp. 1202-1213.

"Unfortunately, extensive gaps in the fossil record of the Lower Carboniferous and Triassic make it very difficult to establish the nature of relationships among Paleozoic tetrapods, or their specific affinities with modern amphibians. An even wider temporal and morphological gap separates modern amphibians from any plausible Permo-Carboniferous ancestors."

Gilbert, Scott., "Morphogenesis of the Turtle Shell: the Development of a Novel Structure in Tetrapod Evolution", *Evolution and Development 3*, March-April 2001, p. 56.

"The turtle shell represents a classic evolutionary problem; the appearance of a major structural adaptation (due to the) instantaneous appearance of this evolutionary novelty. The distinctive morphology of the turtle appears to have arisen suddenly."

Greene, Harry W., Naturalist/Darwinist, with Patricia and Michael Fogden, *"Snakes: The Evolution of Mystery in Nature"*, University of California Press, 2000, p. XII and p. 14.

"We bury snakes and lizards in Class Reptilia, together with turtles- a group with which they haven't shared a common ancestor in more than 200 million years. Traditional classifications often fail to portray evolutionary relationships."

Carroll, Robert L., Naturalist/Darwinist, *"Patterns and Processes of Vertebrate Evolution",* **Cambridge University Press, 1997, pp. 296-297. (2 quotes)**

"The origin of the modern amphibian orders' (and) 'the transition between early tetrapods' are still 'poorly known', along with the origin of many other major groups."

Gould, Stephen Jay., Naturalist/Darwinist, "Eight (or Fewer) Little Piggies", *Natural History,* **vol. 100, no. 1, January 1991, p. 25.**

"Evolutionists at one time claimed that the Seymouria fossil was a transitional form between amphibians and reptiles. According to this scenario, Seymouria was 'the primitive ancestor of reptiles.' However, subsequent fossil discoveries showed that reptiles were living on earth some 30 million years before Seymouria. In the light of this, evolutionists had to put an end to their comments regarding Seymouria. No fossil amphibian seems clearly ancestral to the lineage of fully terrestrial vertebrates (reptiles, birds, and mammals. "

Colbert, Edwin., Naturalist/Darwinist, *"Evolution of the Vertebrates",* **New York: John Wiley & Sons, 1991, pp. 99 and 193.**

"...there is no evidence of any Paleozoic amphibians combining the characteristics that would be expected in a single common ancestor. The oldest known frogs, salamanders, and caecilians are very similar to their living descendants. (There are) no clues in pre-Triassic sediments as to the possible ancestors of the ichthyosaurs (first marine reptiles)."

Carroll, Lewis., Naturalist/Darwinist, *"Problems of the Origin of Reptiles",* **Biological Reviews of the Cambridge Philosophical Society, 1969, vol. 44, p. 393.**

"Unfortunately not a single specimen of an appropriate reptilian ancestor is known prior to the appearance of true reptiles. The absence of such ancestral forms leaves many problems of the amphibian-reptilian transition unanswered."

REPTILE > BIRD

The fossil record contains 165 million years of reptile activity, which generated billions of fossilized reptiles. There should be a clear trail of evidence showing the slow and gradual progression from reptile to bird, if the former evolved into the latter. *'Archaeopteryx'* is the main fossil presented as a transition. It was first found in Bavaria, and the most recent specimen *(number 7 found in 1993)* is the only bird/reptile fossil routinely taught in textbooks. However, many extinct birds, and a few birds living today, have all of Archaeopteryx's characteristics. All of its physical features were fully bird like, and it had no incipient structures. So, archaeopteryx was simply, a bird. DNA tests have also proven that the genes of reptiles and birds can not be related because of significant genetic differences in their finger bones, or "digits". [4, 12, 14]

(Internet Archaeopteryx Photo from: http://www.healthstones.com)

Even more recently there have been reports of "feathered dinosaurs", like *Sinosauropteryx prima*, discovered in China. After further examination, the "feathers" were discovered to be an array of fibers, probably collagen. *Mononykus* was an alleged flightless 'dino-bird' that was later clearly identified as a fleet-footed fossorial *(digging)* theropod. Others, *Protarchaeopteryx robusta*, and *Caudipteryx zoui*, were widely publicized to be new missing links but their 140-150 million year-old dates make them far younger than their supposed ancestors. Most paleontologists consider the valid fossils to be flightless birds similar to the ostrich, not reptile-bird links. One recent "transitional fossil" from China (1999), *Archaeoraptor liaoningensis*, was proven to be a falsified hoax and became an international embarrassment to prestigious journals like National Geographic, who prematurely reported it to be a macroevolutionary fossil. So desperate are Darwinists to find a reptile>bird fossil, or ANY transitional fossil, that they often make quick news releases, only to be embarrassed later. This pattern is all too common and is very bad science. [4, 12, 14]

"Archaeorapter *Liaoningensis: "Fake Dinosaur-bird Ancestor"*, http://www.nwcreation.net/ evolutionfraud.html, March 2009. *(2 quotes)*

> *"The most recent and perhaps the most infamous evolution fraud (see "Feathers For T-Rex?" by Christopher P. Sloan) was committed in China and published in 1999 in the journal National Geographic 196:98-107, November 1999. Dinosaur bones were put together with the bones of a newer species of bird and they tried to pass it off as a very important new evolutionary intermediate."* … *"National Geographic has reached an all-time low for engaging in sensationalistic, unsubstantiated, tabloid journalism' says Storrs L. Olson of the Smithsonian Institution."*

Taylor, Joe and Henry Johnson., Mt. Blanco Fossil Museum and the 1st Institute of Omniology, *http:// www.omniology.com/Joe&Henry.html* , March 2009.

> *"It is obvious to this omniologist that the National Geographic Society passionately embraces, with a biased and dogmatic zeal, the neo-Darwinian (macroevolutionary) world view only!"*

Hickman, C.P., with Roberts and Larson., *"Integrated Principals of Zoology"*, McGraw Hill, NY, 899, 2001, p. 583.

> *"…the fossil record of birds is disappointingly meager."*

Sereno, Paul C., "The Evolution of Dinosaurs", *Science* **284(5423):2137–2147, June 25, 1999, p. 2143.**

"For use in understanding the evolution of vertebrate flight, the early record of pterosaurs and bats is disappointing: their most primitive representatives are fully transformed as capable fliers."

Olson, Storrs., dinosaur expert at the Smithsonian Institute, *Interview***, Oct. 1997.**

"That is the end of it (the debate that birds came from dinosaurs), as far as I am concerned. There is no way that birds and dinosaurs could be directly related."

Feduccia, Alan., Chairman of Biology, University of North Carolina, *Interview***, Oct. 1997.**

"We consider this (digit data) to be unequivocal evidence that birds did not evolve directly from dinosaurs."

Martin, Larry D., "The Barosaurus is No Five-Story-Tall Canary", *Sunday World Herald***, Omaha, Nebraska, 19 January 1992, p. B-17.**

"The theory linking dinosaurs to birds is a pleasant fantasy that some scientists like because it provides a direct entry into a past we otherwise can only guess about. But unless more convincing evidence is uncovered, we must reject it and move on to the next better idea."

Feduccia, Alan., Chairman of Biology, University of North Carolina, *Science***, 259:790-793, 1993.**

"Archaeopteryx probably cannot tell us much about the early origins of feathers and flight in true proto-birds because Archaeopteryx was, in a modern sense, a bird."

"World Book Encyclopedia", **(regarding reptiles becoming birds), Vol. 2, 1982, p. 291.**

"No fossil of any such birdlike reptile has yet been found."

Ostrom, Hohn., "Bird Flight: How Did It Begin?", *American Scientist***, vol. 67, 1979, p. 47.**

"No fossil evidence exists of any pro-avis (bird ancestor). It is a purely hypothetical pre-bird, but one that must have existed."

Swinton, W.E., British Museum of Natural History, *"Biology and Comparative Physiology"***, A. J. Marshall (editor), Vol. 1, New York, 1960, p. 1.**

"The (evolutionary) origin of birds is largely a matter of deduction. There is no fossil evidence of the stages through which the remarkable change from reptile to bird was achieved."

REPTILE *(cold-blooded)* > MAMMAL *(warm-blooded)*:

There are a minimum of 21 major physiological differences separating cold-blooded reptiles and warm-blooded mammals. These complex body changes should have left incredible numbers of transitional fossils, but none have been found. Since no transitional fossils currently exist, Darwinists are forced to assume that mammals descended from one of the more homologically similar reptiles. All 32 orders of mammals *(marsupials, placentals, flying mammals, rodents, marine, primates, etc...)* appear abruptly and fully formed in the fossil record, with no preceding species, and with no transitional fossils between them.

One iconic textbook example that Darwinists promote as one of their best examples of mammalian macroevolution is the "horse" series. However, the animals in this series are simply a group of homologically similar mammals with no transitional fossils. Their fossil 'connection' is so poorly established that most text books and museums have now removed this series from their evolution evidence presentations. Another newly rejected "classic" is Darwin's famous Galapagos finch beak study. While the study of finch beaks does provide an excellent example of micro-genetic variation within a species, it is in no way an example of macroevolution, because it is now known that finch beaks cyclically lengthen and shorten with environmental conditions.

There are currently 4,300 species of mammals currently recognized, plus thousands known to have become extinct. [4, 12, 14] Macroevolutionary transitions should be abundant, but once again, the experts agree that none exist:

Salleh, Anna., Naturalist/Darwinist, "DNA Evidence Settles Debate on Mammal Origins", *ABC Science Online*, 2 February, 2001.
"Humans are closer to rats than they are to any other distant relatives, according to new research which claims to settle a number of hot debates on the origin of mammals. The evolutionary history of the placental mammals has provided grounds for scientific debate as acrimonious as you would find anywhere. Over the past decade, evolutionary biologists seeking to reconstruct the evolutionary relationships of placental mammals using DNA and molecular evidence have often locked horns with scientists using an approach based on careful study of bones, teeth and anatomy. 'Up until now, we've really only had mickey-mouse approaches to studying these questions."

Hoyle, Sir. Fred., Physicist and Astronomy Professor, Cambridge University, *"Mathematics of Evolution"*, Acorn Enterprises: Memphis, TN, 1999, p. 107. *(underlines added)*
"From 1860 onward, the more distant fossil record became a big issue, and over the next two decades, discoveries were made that at first seemed to give support to the theory, particularly the claimed discovery of a well-ordered sequence of fossil horses dating back about 45 million years. Successes like this continue to be emphasized both to students and the public, but usually without the greater failures being mentioned. Horses, according to the theory, should be connected to other orders of mammals, which common mammalian stock should be connected to reptiles, and so on backward through the record. Horses should thus be connected to monkeys and apes, to whales and dolphins, rabbits, bears. ... But such connections have not been found. Each mammalian order can be traced backward for about 60 million years and then, with only one exception the orders vanish without connections to anything at all. The exception is an order of small insect-eating mammal that has been traced backward more than 65 million years..."

Eldredge, Niles., as quoted in: Luther D. Sunderland, *"Darwin's Enigma: Fossils and Other Problems"*, fourth edition, Master Book Publishers, Santee (California), 1988, p. 78.
"I admit that an awful lot of that (imaginary stories) has gotten into the textbooks as though it were true. For instance, the most famous example still on exhibit downstairs (in the American Museum) is the exhibit on horse evolution prepared perhaps 50 years ago. That has been presented as literal truth in textbook after textbook. Now I think that that is lamentable ...".

Lewin, Roger., Naturalist/Darwinist, "Bones of Mammals", *Science*, vol. 212, June 1981, p. 1492.
"The transition to the first mammal, which probably happened in just one or, at most, two lineages, is still an enigma."

Kemp, T. S., Naturalist/Darwinist, *"Mammal-like Reptiles and the Origin of Mammals"*, **New York: Academic Press, 1982, pp. 3 and 583.** *(2 quotes)*
"Gaps at a lower taxonomic level, species and genera, are practically universal in the fossil record of the mammal-like reptiles. In no single adequately documented case is it possible to trace a transition, species by species, from one genus to another."... "Each species of mammal-like reptile that has been found appears suddenly in the fossil record and is not preceded by the species that is directly ancestral to it. It disappears some time later, equally abruptly, without leaving a directly descended species..."

Rensberger, Boyce., *Houston Chronicle*, **5 November 1980, sec. 4, p. 15.**
"The popularly told example of horse evolution, suggesting a gradual sequence of changes from 4-toed fox-sized creatures living nearly 50 million years ago to today's much larger 1-toed horse, has long been known to be wrong. Instead of gradual change, fossils of each intermediate species appear fully distinct, persist unchanged, and then become extinct. Transitional forms are unknown."

Kerkut, Prof. G. A., Dept. of Physiology and Biochemistry, University of Southampton, *"Implications of Evolution"*, **Pergamon Press, London, 1960, p. 144.**
"The evolution of the horse provides one of the keystones in teaching of evolutionary doctrine, though the actual story depends to a large extent upon who is telling it and when the story is being told. In fact, one could easily discuss the evolution of the story of the evolution of the horse!"

LAND MAMMAL > MARINE MAMMALS:

Darwin hypothesized that cold-blooded water vertebrates *(fish)*, must have evolved into warm-blooded marine mammals *(**Cetaceans**, like dolphins/whales)* and that these marine mammals would have later evolved legs as they transitioned onto land to become land mammals. But the fossil record presents two huge problems for Darwinism. Problem one is that land mammals appeared first in the fossil record; Cetaceans appeared much later. Problem two is that there is no chain of transitional fossils linking land and marine mammals. The lack of fossils documenting the evolution of land mammals to Cetaceans is one of the widest and most profound gaps in all of Darwinism. Because Cetaceans are large-boned animals that would be immediately buried in soft preserving sediments, marine mammals have the highest potential of fossilization. They should have left a very clear record of transitional forms, but none exist. The oldest whale fossils in the geologic record show that they were completely aquatic from the first time they appeared. Five creatures, *Mesonychid, Ambulocetus, Pakicetus inachus, Rodhocetus,* and *Prozeuglodon* are typically presented as transitions, but their ages and their lack of incipient physical structures prevent them from being interpreted as true land mammal-whale links. The closest 'relatives' to today's whales are considered to be wolfs, hippos, camels, and pigs. [4, 12, 14]   Most scientists instantly recognize the enormity of that homological fossil gap:

Helsel, Gwen., "Cetacean Evolution", *http://www.pacific.edu/~e-buhals/cetacean*, **2009.**
   "Within the past ten years fossils have been recovered which suggest that members of the order Cetacea are actually descendants of terrestrial animals related to even-toed ungulates, including ruminants, hippos, camels, and pigs (Milinkovitch, 1997). The common ancestor of Cetaceans was actually a land animal that walked on four legs. It was a meat eater that resembled a short-legged wolf with hoof-like claws."

Royal British Columbia Museum Whales On-Line School Program, "Evolution of Whales", *www.royalbcmuseum.bc.ca*, **April 2009.** *(2 quotes)*
   "No one is certain how whales came to exist, but there is fascinating evidence for the evolutionary link between whales and other mammals. ...Scientists now believe that whales evolved from carnivorous land mammals called mesonychids. The huge, furred, wolf-like Andrewsarchus was a mesonychid that lived from 42 to 40 million years ago in the Eocene Epoch."

"The Origin of Marine Mammals", *www.theoryofdarwin.com*, **November 1, 2007.** *(2 quotes)*
   "For evolutionists, the origin of marine mammals has been one of the most difficult issues to explain. In many evolutionist sources, it is asserted that the ancestors of Cetaceans left the land and evolved into marine mammals over a long period of time. Accordingly, marine mammals followed a path contrary to the transition from water to land, and underwent a second evolutionary process, returning to the water. This theory both lacks paleontological evidence and is self-contradictory. Thus, evolutionists have been silenced on this issue for a long time... the scenario was based on evolutionist prejudice, not scientific evidence."

Normile, D., "New Views of the Origins of Mammals", *Science*, **Vol. 281, 7 August, 1998, pp. 774-775.** *(2 quotes)*
   "By some analysis, hippos are the closest living whale relative. ... He (O'Leary) says it is difficult to connect hippos with whales in the fossil record."

Matthews, L. Harrison., Naturalist/Darwinist, *"The Natural History of the Whale"*, **Columbia University Press, New York, 1978, p. 23.**
   "We have no certain knowledge of their origin (Cetaceans), for the earliest known fossils from the Eocene are already, unmistakably, whales, and we can only guess at their evolutionary history by inference."

"Dolphins and Whales", **University of Michigan Press, 1962, p. 17.**
   "We do not possess a single fossil of the transitional forms between the aforementioned land animals and the whales".

Colbert, E. H., Naturalist/Darwinist, *"Evolution of the Vertebrates"*, **1st ed., New York: John Wiley and Sons, 1955, p. 303.**
   "These (marine) mammals must have had an ancient origin, for no intermediate forms are apparent in the fossil record between whales and the ancestral Cretaceous placentals (mammals). Like the bats, the whales appear suddenly in early Tertiary times..."

CHAPTER 8: THE EVOLUTION OF MAN

APES > HUMAN:

The attempt to link apes to humans is probably the most famous and controversial area of Darwinism, but like all of the other transitions previously mentioned, the fossil gap between apes and humans is profound and wide. Apes and humans both appear suddenly and fully formed in the fossil record, with no convincing evolutionary precursors. Linking them together would, of course, require half-ape and half-human transitional fossils with incipient structures. Many skeleton fragments have been presented as potential ape-human *'missing-links'* over the past 90 years. Paleontologists have named this category of fossils: *'hominids'*. Nearly all hominids have been proven to be either outright hoaxes *(like Nebraska Man, Heidelberg Man, Orce Man, Piltdown Man, Peking Man, etc...)*, or they have been clearly classified as ape or human. Skeletons in the *'Australopithecus'* genus are apes, and skeletons in the *'Homo'* genus are human. One key Australopithecus fossil, *"Lucy"*, is currently the most famous and important missing-link, and the science behind her fossil collection and interpretation is hotly debated among paleontologists. [4, 12, 14] New DNA technologies are clarifying and verifying the gap separating apes and humans:

Biggs, Daniel, Lederman, Ortleb, Rillero, Zike, et al., *"Life Science"*, a secondary public school text book published by Glencoe McGraw-Hill, New York, NY, 2009, pp. 171-173. ISBN # 0-07-823695-9, online at www.science.glencoe.com,   *(several quotes).*

> *"The oldest hominid, Lucy, is thought to have walked upright... Scientists have suggested that Homo habilis gave rise to Homo erectus... and they are thought to be ancestors of humans. ... Neanderthals might represent a side branch of human evolution and might not be direct ancestors of modern humans. ...Cro-Magnons are thought to be direct ancestors of early humans."*

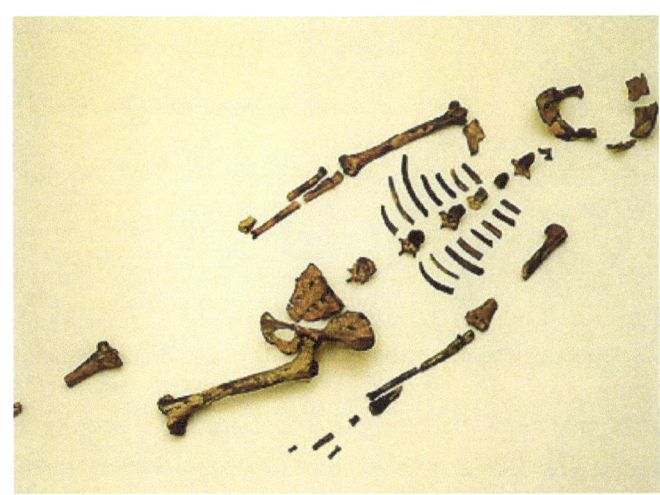

("Lucy" Photo, courtesy of www.foxnews.com)

Israeli Researchers., "Lucy is Not the Direct Ancestor of Humans", *http://www.jpost.com*, April 16, 2007.

> *"Even Lucy, the supposed non-questionable 'missing-link of human evolution' has recently been completely overthrown from her lofty status as irrefutable proof for human evolution. The presence of the morphology in both the latter and Australopithecus afarensis and its absence in modern humans cast doubt on the role of (Lucy) as a common ancestor."*

Oxnard, Charles., Naturalist/Darwinist, Professor of anatomy at the University of Southern California Medical School, "New Fossil Ape May Shake Human Family Tree", *www.news.nationalgeographic*, August 22, 2007.

> *"The Australopithecines (Lucy) known over the last several decades from Olduvai, Sterkfontein, Kromdraai and Makapansgat, are now irrevocably removed from a place in a group any closer to humans than to African apes, and certainly from any place in a direct human lineage."*

White, et al., "Asa Issie, Aramis and the Origin of Australopithicus", *Nature*, vol. 440:883-889, April 13, 2006.

> *"...The origins of Australopithicus (apes) were obscured by a sparse fossil record."*

Gee, Henry., "Return to the Planet of the Apes," *Nature*, Vol. 412, 12 July 2001, p. 131.

> *"Fossil evidence of human evolutionary history is fragmentary and open to various interpretations. Fossil evidence of chimpanzee evolution is absent altogether."*

Willis, Tom., "Lucy Goes to College", *www.rae.org/lucyknee.html#top*, January, 2001.
(question posed to Dr. Donald Johannson who discovered Australopithecus afarensis (Lucy); in reference to the all-important knee joint that 'proves' Lucy walked upright, rather than like apes):

> *"Roy Holt asked, 'How far away from Lucy did you find her knee? It was very difficult, but Johannson did manage to remember that; '...it was found 60 to 70 meters (over 200 feet) lower in the strata, and two to three kilometers (1.24-1.86 miles) away.'"*

Lewontin, Richard., Naturalist/Darwinist and Professor of Zoology and Biology: Harvard University, *"Human Diversity"*, Scientific American Library: New York, NY, 1995, p. 163. *(2 quotes)*

> *"When we consider the remote past, before the origin of the actual species Homo sapien, we are faced with a fragmentary and disconnected fossil record... no fossil hominid species can be established as our direct ancestor."*

Leakey, Richard., Naturalist/Darwinist, (son of Louise Leakey) Director of National Museums of Kenya, *The Weekend Australian*, May 7-8, 1983. (Echoing the scientific criticisms made of his father's fossils, Richard admitted that...)

> *"Lucy's skull was so incomplete that most of it was imagination, made of plaster of paris, thus making it impossible to draw any firm conclusion about what species she belonged to."*

Gliedman, John. "Miracle Mutations", *Science Digest*, vol. 90, February 1982: pp. 90-96.

> *"No fossil, or other physical evidence, directly connects man to ape."*

Watson, Dr. Lyall., "The Water People", *Science Digest*, May 1982, p. 44.

> *"The fossils that decorate our family tree are so scarce that there are still more scientists than specimens. The remarkable fact is that all the physical evidence we have for human evolution can still be placed, with room to spare, inside a single coffin. Modern apes, for instance, seem to have sprung out of nowhere. They have no yesterday, no fossil record."*

Zihlman, A., "False Start of the Human Parade", *Natural History*, vol. 88, 1979, p. 86.

> *"Human nature abhors a vacuum, particularly a genealogical one. There have always been gaps in the fossil record of human evolution but never a shortage of speculative 'missing links'."*

Gould, Stephen Jay., Naturalist/Darwinist, Professor of Geology and Paleontology at Harvard University, "Evolution's Erratic Pace", *Natural History*, May, 1977, p. 13.

> *"The family trees which adorn our textbooks are based on inference, however reasonable, not the evidence of fossils."*

Ager, Derek., Professor of Geology at the Imperial College in London, *"The Nature of the Fossil Record"*, Proc. Geological Association, Vol. 87, 1976, p. 1.

> *"It must be significant that nearly all of the evolutionary stories I learned as a student... have now been 'debunked'."*

Zuckerman, Lord Solly., MA, MD, Dsc. Professor of Anatomy, University of Birmingham, Chief Scientific Advisor, UK, *"Beyond the Ivory Tower"*, Taplinger Publishing Co., 1970, p. 64.

> *"... if man evolved from an apelike creature he did so without leaving a trace of that evolution in the fossil record."*

Simpson, George Gaylord., "The Non-prevalence of Homanoids," *Science*, vol. 143 February 21, 1964, pp. 769-775.

> *"The fossil record shows very clearly that there is no central line leading steadily, in a goal-directed way, from a protozoan to man..."*

APE-HUMAN GENOME SIMILARITIES:

One field of research that is attempting to shed new light on ape-human ancestry is the Human Genome Project. Medical geneticists are mapping the human genome to find the location and function of every one of man's 30,000 genes. This enables doctors to solve medical problems and make medical predictions, but it is also yielding information that is being used by Darwinists to support the claim of ape-human macroevolution. At this point in time, most geneticists agree that the coding of human and chimpanzee DNA shows a similarity of 95-98%, depending on the comparison model being used. Darwinists are suggesting that the genetic similarities are proof of common descent. At first glance this argument sounds entirely reasonable and logical. There is one huge problem with that assertion, however. Similarity of genetic coding also makes an extremely strong intelligent design argument. How so? Here is a simple analogy. Examine the computer software programs that were written to generate two of your utility bills; your phone bill and power bill. They probably have a similarity of 98% even though they were written by two different utility companies. Both computer programs are designed to do the following:

1. Capture and analyze customer usage data, (# of phone calls and # of watts used per day).
2. Categorize those usages according to time of day and day of week.
3. Multiply each usage category by varied and predetermined charge rates.
4. Attach additional fees, like surcharges, state tax, federal tax, etc…
5. Calculate the customer's final bill and apply it to their account.
6. Lastly, both computer programs would: order the billing data to be printed to paper, mail the bill to the address of the customer, and track payments as they are received from the customer.

The reason the two programs have 98% similarity, is because they code for virtually the same functionality. But even though the two billing programs share extremely high similarity in their coding, that does not allow the automatic inference that the programs spontaneously generated from non-intelligent sources, or that the phone bill will 'evolve' into the power bill by random environmental processes over time, or that the phone company's software will someday generate a power bill! Computer programs represent *information* which is conceived, written, and put into action by intelligent agents. Genetic codes are exactly analogous to computer programs, and genetic codes contain even higher programming complexity than man-made billing software. It would only make sense that the genetic coding of two animals as similar as apes and humans would share very similar genetic programming, because organisms having similar physiology would require similar genetics.

Finally, if the software that generated the two utility bills had been written by the same programmer, it would be even more likely for them to have a high degree of similarity. And if *life's* genetic codes were written by the same 'designer', it would make equal sense that ape and human programming would be extremely similar. It is for these reasons that many scientists, myself included, feel that the most interesting question is not; why are organic genetic codes similar… *but who programmed life's complex, ordered, genetic codes?*

Dawkins, Richard., Atheist, Zoologist and Professor for the Public Understanding of Science, Oxford University, *"River out of Eden: A Darwinian View of Life,"* Phoenix: London, 1996, pp. 19-20.
>*"After Watson and Crick, we know that genes themselves, within their minute internal structure, are long strings of pure digital information. What is more, they are truly digital, in the full and strong sense of computers and compact disks, not in the weak sense of the nervous system. The genetic code is not a binary code as in computers, nor an eight-level code as in some telephone systems, but a quaternary code, with four symbols. The machine code of the genes is uncannily computer-like. Apart from differences in jargon, the pages of a molecular biology journal might be interchanged with those of a computer engineering journal."*

Yahyah, Harun., "Darwinist Misrepresentations About the Human Genome Project", *www.harun yahya.com*, February, 2009.
>*"The analysis published in New Scientist, have revealed a 75% similarity between the DNA of nematode worms and man. (New Scientist, 15 May 1999, p. 27) This definitely does not mean that there is only a 25% difference between man and these worms! According to the family tree made by evolutionists, the Chordata phylum in which man is included and Nematoda phylum, were different from each other even 530 million years ago."*

National Human Genome Research Institute, "New Genome Comparison Finds Chimps, Humans Very Similar at the DNA Level", *www.genome.gov*, **Washington, DC, August 31, 2005.**

"(According to) the study's senior author, Robert Waterston, M.D., Ph.D., chair of the Department of Genome Sciences of the University of Washington School of Medicine in Seattle; 'We still do not have in our hands the answer to a most fundamental question: What makes us human? Despite the many similarities found between human and chimp genomes, the researchers emphasized that important differences exist between the genomes, each of which, like most mammalian genomes, contains about 3 billion base pairs. In addition, there are another 5 million sites that differ because of an insertion or deletion in one of the lineages, along with a much smaller number of chromosomal rearrangements. Most of these differences lie in what is believed to be DNA of little or no function. However, as many as 3 million of the differences may lie in crucial protein-coding genes or other functional areas of the genome."

DeWitt, David., *TJ***: 17(1), April 2003.**

"A new report in the Proceedings of the National Academy of Sciences suggests that the common value of >98% similarity of DNA between chimp and humans is incorrect. Roy Britten, author of the study, puts the figure at about 95% when insertions and deletions are included. Importantly, there is much more to these studies than people realize. The >98.5% similarity has been misleading because it depends on what is being compared. There are a number of significant differences that are difficult to quantify. A review by Gagneux and Varki described a list of genetic differences between humans and the great apes. The differences include 'cytogenetic differences, differences in the type and number of repetitive genomic DNA and transposable elements, abundance and distribution of endogenous retroviruses, the presence and extent of allelic polymorphisms, specific gene inactivation events, gene sequence differences, gene duplications, single nucleotide polymorphisms, gene expression differences, and messenger RNA splicing variations. These types of differences are not generally included in calculations of percent DNA similarity."

Wood, Todd Charles Ph.D., Human Genome Expert: Professor at the Center for Origins Research, "The Human Genome", *www.icr.org/article/451*, **May 2001.**

"The size of the human genome is easily misinterpreted, however. One might think that a human might have 25 times as much DNA as a fly because humans are so much larger and more complex. Unfortunately, the amount of DNA in a genome appears to be uncorrelated with biological complexity. For example, the single-celled ciliate, Paramecium caudatum, possesses a genome of 8.6 billion nucleotides, more than twice as big as the human genome. One of the largest known genomes, 670 billion nucleotides, is found in the single celled Amoeba dubia."

Next, let's examine some scientific facts about 'hominid' *(half human- half ape)* fossils that contradict classic Darwinism. The following pages examine contemporary human fossil data and, for the sake of comparison, traditional radiometric dates will be accepted. Readers should keep in mind the previously stated arguments that can, and should, be raised concerning the "selection" of radiometric dates.

TYPICAL COMMERCIAL PRESENTATION OF HOW MAN EVOLVED FROM APE

(What you are never told is...)

Most of these animals lived at the same time...

These animals are not in the exact order found in the geologic column, and several are based only on bone fragments.

So, even though it would not promote the cause of Darwinism, it would be just as scientifically accurate to present the above diagram in the following order:

GENUS: 'AUSTRALOPITHICUS'........... (APES)

All of the fossils classified in the genus **'Australopithicus'**, are currently known to be true apes. Examples include: A. robustus, A. boisei, A. africanus: *(Taung)*, and A. afarensis: *(Lucy)*. Those four are all extinct, and they are universally agreed to be fully primate. Darwinism does not contend that apes turned into humans. It teaches that apes and humans had a common ancestor, and that 'hominid' fossils show the pathway back to that ancestor. There are three major problems with this theory. Problem one is, primates appear suddenly in the fossil record with *no* evolutionary precursors. Problem two is that some fully human fossils appear in the fossil record *BEFORE* Australopithecine and 'hominid' fossils, which means that they lived as contemporaries with them throughout their entire history. Problem three is that even though no common ancestor is seen in the fossil record, textbooks still often refer to some Australopithecines as hominids regardless of the actual fossil evidence. [4, 12, 14]

GENUS: 'HOMO'.........(HUMANS)

All true human fossils are classified in the genus, **'Homo'**, and are subdivided into the following sub-groups: *Homo erectus, Neanderthal, Archaic Homo sapien, and Modern Homo sapien.* During the 20th century, many of these bones were incorrectly misclassified as 'hominids' so many corrections have occurred in the past 2 decades. Here is a summary of what is currently known about each category of human fossils:

HOMO ERECTUS MAN

'Homo erectus' fossils were once considered to be a major hominid species bridging the gap between Australopithecines *(apes)* and Homo sapiens *(humans)*. Most paleontologists now classify Homo erectus and Neanderthal *(see next page)* together as fully human. There are many reasons for interpreting Homo erectus fossils as fully human and well within the range of genetic variability. Their average cranial capacity is 1000cc, and could easily be classified as modern Homo sapien, whose cranial sizes range from 700-2200cc. Homo erectus fossil evidence is quite complete and shows that they lived side by side with true humans for the past 2 million years *(using evolutionary chronology)*. This eliminates the possibility of Homo erectus macroevolving into Homo sapiens. The frontal bone flattening of Homo erectus is seen today in humans with micro-cephalic disease. Slight differences in bone formation could also be the result of rituals and/or diseases that today's surviving primitive tribes still exhibit: head binding, foot binding, rickets, syphilis, etc… They used stone tools, fire, buried their dead, and used red ochre… as did almost every category of fossilized humans. Over half *(62 of 106)* of all Homo erectus fossils are carbon dated at less than 12,000 ya, some even date to just a few hundred years old! This fact also destroys the entire concept of Homo erectus being a hominid link in human macroevolution.

One famous fossil, originally thought to be a H. erectus hominid, is known as "Java Man". In 1892 on the island of Java, Dutch physician Eugene Dubois found a thigh-bone, which for all intents and purposes was like that of modern humans. About a year earlier in roughly the same location he had found a large skull-cap, and later three teeth. These were not necessarily from the same individual because the skullcap and the leg-bone were about 15 meters (50 feet) apart, nevertheless, Dubois published that he had found a 'missing link'. It eventually became widely accepted as such in spite of the fact that a leading authority had identified two of the teeth as those of an orangutan, and the other tooth as human. Java man was trumpeted around the world as indisputable proof of human evolution. Textbooks and magazines were filled with hypothetical reconstructions of Java man and it was given the impressive-sounding scientific name of *Pithecanthropus erectus,* or (*'erect ape-man'*). According to: Darwinist Stephen Jay Gould, in: "Men of the Thirty-Third Division", *Natural History*, April, 1990, pp. 12–24: *Pithecanthropus was not a man, but a gigantic genus allied to the gibbons."* Homo erectus was once thought to be a contemporary of another ancient fossil group called **'Homo habilis'**. Habilis was also once classified as part of the hominid transitional line, but like Java Man, it was later discovered to be a flawed taxon. Habilis' fossils were eventually proven to be an incorrect mixture of true human and nonhuman fossils. Either way, by date or by fossil evidence, Homo habilis and Java Man have now been removed as a hominid species. [4, 12, 14]

NEANDERTHAL MAN

Neanderthal fossils were also once considered to be a transitional hominid by many paleontologists, but they are now known to be an isolated branch of modern humans. The most recent carbon dating suggests that Neanderthals lived as recently as 28,000 years ago *(Vindija cave, Croatia)*. They have a **LARGER** average brain size than modern humans, 1620cc Vs 1450cc. They made and used complex tools, built sophisticated dwelling structures, buried their dead, etc… Their skeletons are only superficially different than modern Homo sapiens; lower/wider cranium, heavier brow ridges, larger brain sizes, a more pointed rear of skull, larger/longer facial bones, and a weaker, more rounded chin. All of these characteristics can be exhibited in modern humans and could be attributed to human genetic variation and tribal isolation. Their somewhat unique facial, skull, and jaw morphology could have also resulted from the unique stresses their jaws and teeth were subjected to when used as oral tools. There is also a very strong forensic argument that they were a "tribe" of modern humans who were racked with malnutrition, severe rickets, arthritis, and/or possibly syphilis. [4, 12, 14]

ARCHAIC HOMO SAPIEN MAN

Roughly 50 fossil fragments fit this group, the most famous and complete is called "Rhodesian Man". Cranial capacity is within the range of modern man at 1100-1300cc. They had heavier ridges over the eyes, a more rounded rear skull, longer face with jutting jaw, and post cranial bones indistinguishable from modern humans. Most anthropologists now consider Archaic Homo sapien to be a fully modern human, and it also lived as a contemporary along side modern humans. Its slight physiological differences were probably the result of existing human genetic diversity, disease, and/or tribal isolation. There is evidence that they had a relatively high degree of civilization and technology. [4, 12, 14]

MODERN HOMO SAPIEN MAN

Darwinists insist that apes and hominids evolved millions of years before modern man, and that modern man has only been present on earth for a little more than one million years. The problem with that theory is, several modern human fossils **PRECEDE** *all "homonid" fossils* by radiometric dating. This fact adds further destruction to the macroevolution theory. The Laetoli footprints in Tanzania are fully human and date back to 3.7 mya using K-Ar radiometric dating. The Kanapoi (KP271) humerous bone from Kenya is also a fully modern human fossil that dates to 4.5 mya. Most macroevolutionists agree that their carbon dates are correct, but choose to disregard these fossils simply because they are "too old" to fit the classic evolutionary scenario. Modern human cranial capacities range from 700cc-2200cc, and this range encompasses all known hominid cranial sizes. Currently documented human genetic variations demonstrate body shapes and sizes that include all hominid characteristics. Forensic evidence also shows that disease deformities of modern humans could easily explain the minor physiological oddities previously thought to be hominid characteristics. *'Cro-Magnon'* man, also once classified as a hominid, is now considered by most scientists to be an isolated tribal group of modern Homo sapien. [4, 12, 14]

SUMMARY OF HUMAN FOSSILS TO DATE:

1. Apes appear in the fossil record suddenly, with no evolutionary precursor. They stand separate and distinct from true humans.

2. Fossils suggest that true humans have been on the scene for 4.5 million years, the same as hominids, but the oldest fossils have been discarded by evolutionists as 'bad dates'.

3. Using radiometric dates, Homo erectus, Neanderthal, Cro-Magnon, Archaic Homo sapiens and modern Homo sapiens all lived as contemporaries, side by side.

4. The physiological differences between all human subgroups are slight and could all be attributed to human genetic variation, tribal isolation, malnutrition, and disease.

5. Even if evolutionary fossil dates are accepted without challenge, no Australopithecus fossils could have served as evolutionary ancestors to humans because of their contemporary ages.

PROBLEMS WITH STUDYING ANCIENT HUMAN FOSSILS:

1. "Hominid" fossils are seldom available for independent cross-examination. Only those in the Darwinian inner circle are permitted access. One reason for this is their fragility and irreplaceability. Another reason is to control who examines them, and how the fossils are described. Seldom are 'creationists' allowed access, thus there are few serious critiques of Darwinian fossil reports.

2. Only plaster cast replicas are available for critical examination. They are usually poor quality reproductions, not recommended for serious paleoanthropologic research.

3. Many fossils have been selectively excluded when they did not fit well into the pre-conceived macroevolutionary time scheme, and this filter destroys objectivity.

4. True human fossils are often *downgraded* to hominid status if their radiocarbon dates do not fit pre-conceived macroevolutionary time lines.

5. True ape fossils are often *upgraded* to hominid status if their radiocarbon dates do not fit the pre-conceived macroevolutionary time lines.

6. Arranging a series of fossils in a homological sequence is no proof that they evolved. It is just as possible that they were created by a designer using similar design.

7. There are an estimated 6000 catalogued human fossils, but most are fragments, and very few are in the sought after 'ancient' category.

CHAPTER 9: DO TODAY'S EVOLUTION TEXT BOOKS CONFIRM THAT NO TRANSITIONAL FOSSILS EXIST?

Let's examine a typical biology text book currently used by millions of secondary level public school students across the United States. If you look at the specific wording, *(caps were added for emphasis)*, the authors openly admit that there are no transitional fossils to support Darwinian macroevolution. All of the following quotes come from: ***"Life Science"*, published by Glencoe McGraw-Hill, New York, NY, ISBN # 0-07-823695-9, authors: Biggs, Daniel, Lederman, Ortleb, Rillero, Zike, et al., available online at www.science. glencoe.com, 2009.**

Concerning the origin of life; 'bacteria': (The book offers no evidence concerning the origin of first life beyond the fact that biogenesis *(p. 19)* and the cell theory *(p. 51)* both agree that life comes only from similar life. Spontaneous generation, or abiogenesis, *(p. 19)* is completely *REJECTED* by the text book as a life creating force.)

Concerning single-celled protozoa: *(p. 213) "Scientists HYPOTHESIZE that the common ancestor of most protists was a one-celled organism with a nucleus and other cellular structures. However, they COULD HAVE HAD a different ancestor…"*

Concerning fungi: *(p. 230) "Scientists HYPOTHESIZE that early fungi attached themselves to the roots of early plants…"* (The book makes no mention of their origin.)

Concerning the major plant lines: *(p. 243) "…plant ancestors were PROBABLY ancient green algae… this has led scientists to THINK that plants and green algae have a common ancestor. Scientists HYPOTHESIZE that some of these kinds of plants evolved into the plants that exist today. Cone-bearing plants such as pines PROBABLY evolved… however the exact origin of flowering plants IS NOT KNOWN."*

Concerning sponges (Porifera): *(p. 340) "…sponges have little (homology) in common with other animals, so MANY scientists have concluded that sponges PROBABLY evolved separately from all other animals.*

Concerning coelenterates (Cnidarians): *(p. 345) "Scientists HYPOTHESIZE that the medusa body was the first form of cnidarians."*

Concerning flatworms (Platyhelminthes): *(p. 350) "Because of the LIMITED FOSSIL EVIDENCE, the evolution of flatworms is UNCERTAIN. SOME scientists HYPOTHESIZE that flatworms and cnidarians MIGHT HAVE HAD a common ancestor."*

Concerning roundworms (Nematodes): *(p. 353) "It is still UNCLEAR how roundworms fit into the evolution of animals."*

Concerning segmented worms (Annelids): *(p. 373) "SOME SCIENTISTS HYPOTHESIZE that segmented worms evolved in the sea. The fossil record for segmented worms is LIMITED…"*

Concerning Mollusks: *(p. 367) "Mollusks have changed very little from their ancestors… today's MOLLUSKS are descendants of ANCIENT MOLLUSKS."*

Concerning Arthropods: *(p.382) "Scientists HYPOTHESIZE that arthropods PROBABLY EVOLVED from an ancestor of segmented worms."*

Concerning Echinoderms: *(p. 387) "The earliest echinoderms MIGHT HAVE HAD bilateral symmetry as adults, and MIGHT HAVE BEEN attached to the ocean floor… Scientists HYPOTHESIZE that echinoderms MORE CLOSELY RESEMBLE animals with backbones than any other group of invertebrates."*

Concerning fish: *(p. 410) "Today's bony fish are PROBABLY DESCENDED from the first jawed fish… Modern sharks/rays are PROBABLY DESCENDED from placoderms."*

Concerning amphibians: *(p. 415) "Amphibians are THOUGHT TO HAVE EVOLVED from these lobe-finned fish…"*

Concerning birds: *(p. 439) "SOME SCIENTISTS HYPOTHESIZE that birds developed from reptiles millions of years ago."*

Concerning primates: (Nothing mentioned because there are *no* known pre-primate fossils.)

Concerning humans: *(p. 171-173) "The oldest hominid, Lucy, is THOUGHT to have walked upright. Scientists have SUGGESTED that Homo habilis gave rise to Homo erectus… and they are THOUGHT TO BE ancestors of humans. Neanderthals MIGHT represent a side branch of human evolution and MIGHT NOT BE direct ancestors of modern humans. Cro-Magnons are THOUGHT TO BE direct ancestors of early humans."*

Conclusion: Reasonable people surely must ask, do the above text book quotes present a solid evidential case for macroevolutionary fossil progression over time? I give these text book authors high marks for honesty and accuracy, because their hollow, vague, and unsupported fossil statements are universal in all evolution curricula. Text book authors choose every word and every sentence quite carefully. If they actually had any macroevolutionary fossils, they would clearly articulate the transitional history leading from one life group to another. The only type of evolution that can actually be proven is inter-species microevolution, and this was the only type of evolution that Charles Darwin documented in his journals. These text book authors agree with, and corroborate statements from previously quoted scientists throughout this book; there was, and still is, absolutely no macroevolutionary fossil evidence linking different kingdoms and phyla. Without tangible physical evidence, the Darwinian Theory collapses to the level of a philosophy, or a religion. So, how do Darwinists attempt to *'fill'* this evidence vacuum? They simply create a new version of Darwinism that *expects* no connecting fossils, and it is called: ***"punctuated equilibrium"***.

CHAPTER 10: PUNCTUATED EQUILIBRIUM

The fossil record clearly shows the abrupt and simultaneous appearance of biological kingdoms and phyla at the Cambrian layer, with no evidence whatsoever of any transitional fossils. There should be many times more transitional fossils than actual living species if traditional Darwinian theories were true. This universal lack of transitional fossils has led to the new theory of punctuated equilibrium. This variation of Darwinism says life mutated, not gradually as Darwin suggested, but in massive leaps leaving no intermediate fossils. For example: a fertilized fish egg is mutated, birth defected, and born as a baby alligator instead of a minnow. This would create no transitional fossils. **Thus, the new proof of Darwinism is its *total lack of evidence.*** The mathematical probability problems increase geometrically, however, when one considers the odds of multiple, beneficial, physiological defects happening suddenly and simultaneously to both males and females, rather than singly and gradually. Most scientists reject evolution via these genetic "Hopeful Monsters". This *admission* and *prediction* of missing fossil evidence further down-grades the Darwinian theory to, essentially, a religion.

QUESTION: *Is it common for widely held scientific theories to later collapse?* YES, many theories, and even *LAWS* have been accepted and later found to be false: the flat earth theory, blood-letting theory, spontaneous generation, the Ptolemaic geocentric theory *(Earth as center of the solar system)*, and more recently... the atomic *Law* of Parity, plus others.

QUESTION: *If there is no hard evidence for Darwinian evolution, and fossil evidence speaks against it, why is it by far the most dominant theory in science textbooks today?* It is my opinion that:

1. Textbook companies have eliminated any potentially religious content since the 1960's due to legal intimidation and the economic fear of losing business with public schools. Intelligent design arguments, while strengthening, are still very risky to text book sales. Also, Naturalists control most textbook authorship and refuse to print a balanced presentation of "evolution".

2. Gallup Polls have consistently shown that the majority of collegiate scientists are atheists. Atheistic scientists openly profess, and believe only in naturalism, and it makes sense that they consider Darwinism alone, as the only valid origins explanation.

3. Scientists who "defect" from the Darwinist camp currently suffer enormous peer ridicule and persecution in the U.S. Many fear for their reputations as well as their jobs. As one immigrating Chinese scientist, Dr. Jun Yaun, recently noted: *"In China we can criticize Darwin but not the government. In the United States you can criticize the government, but not Darwin!"* This academic persecution is exposed in the recent documentary: *"Expelled: No Intelligence Allowed"*, produced by Premise Media Corporation, 2008. [9]

4. Many nationally known groups have taught Darwinism as a proven truth for so many decades, *(like National Geographic, college biology departments, textbook companies, etc...)* that their pride of reputation often rejects the possibility that the theory has been thoroughly debunked. As commonly stated: *"Protecting Darwinism beats rejecting Darwinism!"*

5. Most people are unable to argue the complicated topics of genetics, paleontology, geology, and thermodynamics with scientists. This intimidation greatly limits critical discussion.

6. Most people falsely believe that all scientists are impartial with scientific data, and that they do not let their personal beliefs affect their professional theories.

7. Ample government funding is available for Darwinian research grants. Profit often overshadows truth.

8. There is still much money and fame in making, and faking, evolution fossil discoveries and/or theories.

Conclusion: Increasing numbers of top research scientists feel that punctuated equilibrium is the final, fatal admission by Darwinists of the thoroughly proven reality that there are no linking fossils connecting any of the major phyla of organisms. This final deathblow to Darwinism has caused more and more scientists and educators to abandon the theory altogether in search of scientific models that can better explain the creation of time, space, matter, and organic life.

Foster, Julie., "The Evolving Darwin Debate: Scientists Urge Academic Freedom to Teach Both Sides of the Issue", *2009 WorldNetDaily.com*, p. 1.
> *"… more than 50 Ohio scientists issued a statement this week supporting academic freedom to teach arguments for and against Darwin's theory of evolution. Dr. Robert DiSilvestro, a professor at Ohio State and statement signatory, believes many pro-evolution scientists have not given Darwin's theory enough critical thought. "As a scientist who has been following this debate closely, I think that a valid scientific challenge has been mounted to Darwinian orthodoxy on evolution. There are good scientific reasons to question many currently accepted ideas in this area," he said. "The more this controversy rages, the more our colleagues start to investigate the scientific issues," commented DiSilvestro. "This has caused more scientists to publicly support our statement."*

Miller, Jon D., and Scott, Eugenie C. & Okamoto Shinji., "Public Acceptance of Evolution", *Science*, vol. 313, 2006, p. 765.
> *"The percentage of people in the country who accept the idea of evolution has declined from 45% in 1985 to 40% in 2005. Meanwhile, the fraction of Americans unsure about evolution has soared from 7% in 1985 to 21% last year (2005)."*

Begley, Sharon., 3 quotes from: "Science Finds God", *Newsweek*, July 20, 1998, pp. 1-3.
(says **Allen Sandage**, world-renowned, award winning astronomer,):
> *"It was my science that drove me to the conclusion that the world is much more complicated than can be explained by science."*

(says **John Polkinghorne**, retired, distinguished physicist from Cambridge University):
> *"When you realize that the laws of nature must be incredibly finely tuned to produce the universe we see, that conspires to plant the idea that the universe did not just happen, but that there must be a purpose behind it. For me, the fundamental component of belief in God is that there is a mind and a purpose behind the universe."*

(says **Charles Townes**, who shared the 1964 Nobel Prize in Physics for inventing the laser):

"Many have a feeling that somehow, intelligence must have been involved in the laws of the universe. As a religious person, I strongly sense the presence and actions of a creative being far beyond myself and yet always personal and close by … somehow intelligence must have been involved in the laws of the universe."

Tipler, Frank., professor of mathematical physics, *"The Physics of Christianity"*, New York: Doubleday, 1994.

"When I began my career as a cosmologist some 20 years ago, I was a convinced atheist. I never in my wildest dreams imagined that one day I would be writing a book purporting to show that the central claims of Judeo-Christian theology are in fact true, that these claims are straightforward deductions of the laws of physics as we now understand them. I have been forced into these conclusions by the inexorable logic of my own special branch of physics."

Bethell, Tom., philosopher, *"The American Spectator"*, July 1994, p. 17.

"Evolution is perhaps the most jealously guarded dogma of the American public philosophy. Any sign of serious resistance to it has encountered fierce hostility in the past, and it will not be abandoned without a tremendous fight. The gold standard could go, Saigon abandoned, the Constitution itself slyly junked. But Darwinism will be defended to the bitter end."

Bird, W. R., *"The Origin of Species Revisited"*, vols. I and II, New York: Philosophical Library, 1989, p. 113.

"How interesting, indeed, that evolutionists might think that the evidence for creationism is more compelling to students than the evidence for evolution, or that the teachers of biology are incapable of presenting evolution convincingly, the solution of which is suppression of creationism. I think the better solution is to let creationism and evolutionism, fight it out in the science classrooms everywhere."

Huxley, Thomas H., world-renowned evolutionist, *"The Life and Letters of Thomas Henry Huxley"*, vol. I, ed. L. Huxley (Macmillan, 1903), p. 241.

"…'creation', in the ordinary sense of the word is perfectly conceivable. I find no difficulty in conceiving that, at some former period, this universe was not in existence, and that it made its appearance in six days, or instantaneously, if that is preferred, in consequence of the volition of some preexisting Being. Then as now, the so-called "a priori" arguments against Theism and, given a Deity, against the possibility of creative acts, appeared to me to be devoid of reasonable foundations."

Bird, W. R., *"The Origin of Species Revisited"*, vols. I and II, New York: Philosophical Library, 1989, p. 113.

"Teachers and school boards in public schools are already free under the Constitution of the United States to teach about supernatural origins if they wish in their science classes."

Alexander, R. D., Professor of Zoology at the University of Michigan, *"Evolution Versus Creationism: The Public Education Controversy"*, ed. J. P. Zetterberg, Phoenix: Oryx Press, 1983 p. 91.

"No teacher should be dismayed at efforts to present creation as an alternative to evolution in biology courses; indeed, at this moment creation is the only alternative to evolution. Not only is this worth mentioning, but a comparison of the two alternatives can be an excellent exercise in logic and reason. Our primary goal as

educators should be to teach students to think, and such a comparison, particularly because it concerns an issue in which many have special interests, or are even emotionally involved, may accomplish that purpose better than most other (scientific issues)."

CHAPTER 11: SUMMARIZING CONCLUSIONS

DARWINISM SCIENTIFICALLY *FAILS* ON EVERY MAJOR INTELLECTUAL AVENUE:

1. It predicts that matter and energy is either self-created, or is eternal, ***which all science laws agree are clearly disproven.***

2. It predicts that matter continuously and randomly self-organizes and moves from disorder to order, ***which goes against all observed scientific evidence to date as well as the second law of thermodynamics.***

3. It automatically rejects the possibility of an external intelligent designer, even though living systems exhibit incredibly high levels of design evidence. ***These same scientists accept much lower design "thresholds" when trying to determine the presence of external design in non-living systems.***

4. It predicts that base elements must randomly unite into complex sequences, which then form complex organic molecules like amino acids, proteins, lipids, DNA, RNA, etc... ***This has never been observed, and the odds of this occurring by chance have been thoroughly rejected, both scientifically and mathematically.***

5. It predicts that these highly complex molecules then randomly unite into even higher and more complex sequences to form cellular organelles, which must immediately and spontaneously self-organize into living cells. ***Science has thoroughly rejected all spontaneous generation (abiogenesis) avenues as being capable of creating life.***

6. It predicts that the first spontaneously generated cell's DNA code (which has repeating, orderly patterns and highly specified complexity) was formed from chaotic disorder by random chance, and then macroevolved into all of the millions of species that have ever lived on earth. ***This goes against all current observations, biological laws, and fossil evidence.***

7. It predicts that these "primitive" genetics randomly increased in complexity by mutation, ***which has never been observed, and is agreed by most geneticists and Darwinists to be the worst vehicle for increasing a genetic code's order, complexity, and functionality.***

8. It accepts highly questionable earth dating methods, yielding an age of 4.6 billion years-old earth, ***but rejects dozens of equally scientific and reliable dating techniques*** simply because they yield earth ages far too young to fit the preconceived macroevolutionary theory.

9. ***It continues to promote thoroughly disproved macroevolutionary "evidences"*** like homology vestigial organs, spontaneous generation, and embryonic recapitulation.

10. It originally predicted that macroevolution would be traced in the fossil record by billions of intermediate fossils containing incipient structures. ***However, no true series of transitional fossils with incipient structures has ever been found connecting the biological kingdoms.***

11. It currently predicts that massive mutations happened "quickly", macroevolving life forms without leaving the expected transitional fossils (punctuated equilibrium). ***The problems with this theory are greater than all of the above, and its proof, is its lack of any evidence.***

DESIGN THEORIES ARE IN AGREEMENT WITH TODAY'S SCIENTIFIC EVIDENCE:

1. It predicts that matter and energy cannot self-create by any naturalistic process, but did have a specific moment of creation, ***which is universally agreed upon by all laws of science.***

2. It predicts that all universe energy must have been "wound up" at some eventful moment in the past since it is now "unwinding", ***which is universally agreed upon by today's scientists and physics laws.***

3. It predicts that chaotic disorder cannot assemble into highly complex order without intelligent intervention, ***which agrees with what all fields of science have observed to date.***

4. It predicts that the simultaneous appearance of the major life groups at the Cambrian layer could not occur by any 'spontaneous chemical evolution' process, ***which is agreed upon by most of today's scientists.***

5. It predicts clear and consistent fossil "gaps" between major taxonomic groups, ***which the fossil record confirms.***

6. It predicts that only small microvariational changes should occur within gene pools due to genetic variation and natural selection, ***which the fossil record confirms.***

7. It predicts decay and extinction of life groups over time, ***which the fossil record confirms.***

8. It predicts that inanimate elemental matter, or "primordial soup", cannot spontaneously generate into living organisms, ***which all fields of science and mathematics thoroughly confirm.***

9. It predicts that living cells can only originate from similar pre-existing cells, ***which the Modern Cell Theory and all biological observations confirm.***

10. It predicts that mutation is destructive to a complex, ordered genetic code, and that it cannot cause any significant increase in the quantity and/or complexity of a gene pool, ***which genetic research and observations to date confirms.***

11. It predicts that natural selection, variation, recombination, migration and isolation can cause only microvariation within a gene pool, because they add ***no new quantity of genes*** and they cannot cause macroevolutionary changes, ***which observations to date confirm.***

12. It recognizes that homology, vestigial organs, and embryology are ***completely discredited evidences for macro and microevolutionary change.***

13. It recognizes the unscientific assumptions inherent in all modern attempts to date the ancient Earth, ***so it equally considers all Earth dating methods that are based on scientific measurements.***

14. It allows for the possibility of external intelligent design ***due to the incredibly high levels of design evidences seen in the fields of biochemistry and anatomy.***

===

In conclusion, I completely understand why my scientific colleagues don't want religious groups covertly hijacking science curricula. But I am confused as to why any scientist would unquestioningly stick with Darwin's theory *and* persecute researchers and educators who dare to discuss the **SCIENTIFIC** arguments against it. For those who still maintain that Darwinism is proven science, and that it is an uncontestable theory, I have to ask… why? The evidential problems with Darwinism are so deep, serious, and obvious, that I'm left to wonder how any thinking scientist could ignore the litany of deathblows leveled against it. I can only think of three possibilities. Either:

1. They are incompetent in the arena of true scientific inquiry,
2. They are so dogmatic about Naturalism that they stick with Darwinism because it is the only theory currently attempting to reconcile their atheistic world-view, or…
3. They have only been taught the evidences **FOR** Darwinian evolution, and have never studied any of the scientific arguments **AGAINST** the theory.

I know the third scenario was once the case for me. After studying the full evidential picture, I became completely convinced that Darwinism was a hollow shell, and that it is as dead as the flat earth theory. In my professional opinion, some type of intelligent causal agent HAS to be behind the creation of time, space, matter, energy, and organic life. In the future, I will continue to follow this topic with some slight interest, but I no longer see any reasons why Darwinism should be given any serious scientific consideration. Where the evolution debate will head in the coming years will be an interesting question, because Darwinists have painted themselves into a corner, scientifically. Most of them are beginning to recognize that atheists like Antony Flew and Richard Dawkins are right; matter must have had a super-natural origin, and life must have been engineered by some type of intelligence. But like Dawkins, many naturalists continue to insist that life's designer and creator must have been an organic, intelligent extraterrestrial. Dawkins' recent and stunning admission supporting intelligent design is worth re-reading one final time:

> *"It (intelligent design) could come about in the following way. It could be that at some earlier time, somewhere in the universe, a civilization evolved by some type of Darwinian means to a very, very high level of technology, and designed a form of life that they seeded onto, perhaps, this planet. Now, that is a possibility, and an intriguing possibility. And I suppose it's possible that you might find evidence for that, if you look at the details of biochemistry and molecular biology. You might*

find a signature, of some sort of designer. And that designer could well be a higher intelligence from elsewhere in the universe, but it (that original intelligence) couldn't have first jumped into existence spontaneously." [9]

I hope this book succeeded in presenting the scientific arguments against Darwinism in a fair, accurate, and professional manner. I hope that it was interesting and useful to you, no matter what your age and academic level. The question of where we came from may be the most important and fascinating science question that anyone can ask; it actually opens up an entirely new field of science. This book may not prove where humans came from, but it should confirm where we did **not** come from. And if it were possible to show Charles Darwin the modern fossil record and all of the high-tech scientific evidences gathered during the 150 years since his death, I feel confident that he would agree.

David R. Browning

REFERENCES

[1] Austin, Steven A. *"Grand Canyon: Monument to Disaster"*, ICR, 1994.

[2] Behe, Michael J., *"Darwin's Black Box: The Biochemical Challenge to Evolution"*, New York: The Free Press, 1996.

[3] Behe, Michael J., *"Intelligent Design Theory as a Tool for Analyzing Biochemical Systems"*, Mere Creation: Science, Faith and Intelligent Design, Downers Grove, Il., Intervarsity Press, 1998.

[4] Biggs, Daniel, Lederman, Ortleb, Rillero, Zike, et al., *"Life Science"*, (public school text book) by Glencoe McGraw-Hill, New York, NY, online at www.science.glencoe.com, 2009.

[5] Bliss, Richard Ed. D., *"Origins: Creation or Evolution"*, Master Books, 1988.

[6] Cohen, I., *"Darwin Was Wrong; A Study In Probability"*, New Research Publications, 1984.

[7] Dembski, William A., *"The Design Revolution: Answering the Toughest Questions about Intelligent Design"*, Downer's Grove, Illinois: InterVarsity Press, 2004.

[8] Denton, Michael, PhD., *"Evolution: A Theory in Crisis"*, Adler/Adler, Bethesda, MD, 1985.

[9] *"Expelled: No Intelligence Allowed"*, Premise Media Corporation, 2008.

[10] Flew, Antony, *"There is A God: How the World's Most Notorious Atheist Changed His Mind"*, Harper-Collins Publishers, 2007.

[11] Gert, Werner., *"In the Beginning Was Information"*, Master Books, 2006.

[12] Gish, Duane, Ph.D., *"Evolution: The Fossils Still Say No"*, Master Books, 1995.

[13] Johnson, Phillip., *"Defeating Darwinism"*, InterVarsity Press, 1997.

[14] Lubenow, Marvin L., *"Bones of Contention"*, Michigan: Baker Books, 1992.

[15] McDowell, Josh., *"Evidence That Demands a Verdict"*, Thomas Nelson Publishers, 1979.

[16] Morris, Henry., *"That Their Words May Be Used Against Them"*, ICR, San Diego, 1997.

[17] Morris, John D. Ph. D., *"The Young Earth"*, Master Books, 1999.

[18] Myer, Steven C., et al., *"Science and Evidence of Design in the Universe"*, Ignatius Press, 2000.

[19] Ross, Hugh. *"The Creator and the Cosmos"*, Colorado: NavPress Publishing Group, 1993.

[20] Sanford, Dr. John., *"Genetic Entropy and the Mystery of the Genome"*, Ivan Press, 2005.

[21] Sarfati, Jonathan Ph.D., *"Refuting Evolution, A Response to the National Academy of Science: Teaching about Evolution and the Nature of Science"*, NY: The Free Press, 1996.

[22] Spetner, Lee Ph. D., *"Not By Chance: Shattering the Modern Theory of Evolution"*, New York, Judaica Press, 1997.

[23] Sutherland, Luther., *"Darwin's Enigma"*, Master Books, 1998.

[24] Wells, Jonathan, *"Icons of Evolution: Science or Myth"*, Regnery Publishing Inc., Washington, D.C., 2000.

[25] Werner, Dr. Carl., *"Evolution: The Grand Experiment"*, New Leaf Press, Green Forest Ark., 2007.

[26] WWW.creationscience.com

[27] WWW.Darwinism-Watch.com

[28] WWW.ICR.org

[29] WWW.REASONS.org

[30] WWW.trueorigin.ocg

If you are ever denied the right of free speech in a public school classroom, contact one of the following law firms.

LEGAL WEBSITE LINKS:

www.rutherford.org   (The Rutherford Institute… defends civil and religious rights, free of charge)

www.aclj.org   (The American Center for Law and Justice… also defends civil and religious rights, free of charge)

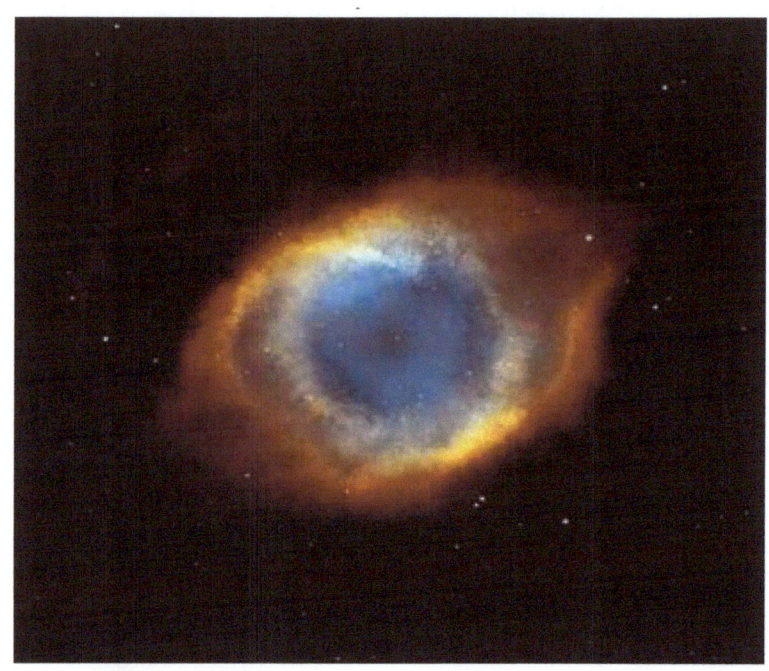

Hubble photo, named: "The Eye of God"

NOTES

NOTES

NOTES

NOTES

www.ingramcontent.com/pod-product-compliance
Lightning Source LLC
Chambersburg PA
CBHW051020180526
45172CB00002B/413